PALGRAVE MACMILLAN'S
CRITICAL STUDIES IN GENDER, SEXUALITY, AND CULTURE

Highlighting the work taking place at the crossroads of sociology, sexuality studies, gender studies, cultural studies, and performance studies, this series offers a platform for scholars pushing the boundaries of gender and sexuality studies substantively, theoretically, and stylistically. The authors draw on insights from diverse scholarship and research in popular culture, ethnography, history, cinema, religion, performance, new media studies, and technoscience studies to render visible the complex manner in which gender and sexuality intersect and can, at times, create tensions and fissures between one another. Encouraging breadth in terms of both scope and theme, the series editors seek works that explore the multifaceted domain of gender and sexuality in a manner that challenges the taken-for-granted. On one hand, the series foregrounds the pleasure, pain, politics, and aesthetics at the nexus of sexual practice and gendered expression. On the other, it explores new sites for the expression of gender and sexuality, the new geographies of intimacy being constituted at both the local and global scales.

Series Editors:

PATRICIA T. CLOUGH is Professor of Sociology and Women's Studies at Queens College and The Graduate Center, CUNY. Clough is on the editorial boards of *Women's Studies Quarterly, Body and Society, Subjectivity, Cultural Studies/Critical Method, Qualitative Inquiry,* and *Women and Performance.* Clough is the coeditor of *Beyond Biopolitics: Essays in the Governance of Life and Death* (with Craig Willse, 2011); author of *The Affective Turn: Theorizing the Social* (with Jean Halley, 2007); *Autoaffection: Unconscious Thought in the Age of Teletechnology* (2000); *The End(s)of Ethnography: From Realism to Social Criticism* (1998); *Feminist Thought: Desire, Power and Academic Discourse* (1994); *The End(s) of Ethnography: From Realism to Social Criticism* (1992).

R. DANIELLE EGAN is Professor and Chair of the Gender and Sexuality Studies Program at St. Lawrence University. Egan is the author of *Dancing for Dollars and Paying for Love: The Relationships between Exotic Dancers and their Regulars* (2006) and coauthor of *Theorizing the Sexual Child in Modernity* (with Gail Hawkes, 2010), both with Palgrave Macmillan. She is also the coeditor of *Flesh for Fantasy: Producing and Consuming Exotic Dance* (with Katherine Frank and Merri Lisa Johnson, 2006). She is on the editorial board of *Sexuality and Culture.*

Titles:

Antarctica as Cultural Critique

The Gendered Politics of Scientific Exploration and Climate Change

Elena Glasberg

First published in 2012 by
PALGRAVE MACMILLAN®
in the United States—a division of St. Martin's Press LLC,
175 Fifth Avenue, New York, NY 10010.

Where this book is distributed in the UK, Europe and the rest of the world,
this is by Palgrave Macmillan, a division of Macmillan Publishers Limited,
registered in England, company number 785998, of Houndmills,
Basingstoke, Hampshire RG21 6XS.

Palgrave Macmillan is the global academic imprint of the above companies
and has companies and representatives throughout the world.

Palgrave® and Macmillan® are registered trademarks in the United States,
the United Kingdom, Europe and other countries.

ISBN: 978–0–230–11687–0

Library of Congress Cataloging-in-Publication Data

Glasberg, Elena.
    Antarctica as cultural critique : the gendered politics of scientific
exploration and climate change / Elena Glasberg.
        p. cm.—(Critical studies in gender, sexuality, and culture)
    ISBN 978–0–230–11687–0 (hardback)
    1. Antarctica—Discovery and exploration. 2. Climatic changes—
History. I. Title.

G860.G56 2012
919.89—dc23   2012013162

A catalogue record of the book is available from the British Library.

Design by Newgen Imaging Systems (P) Ltd., Chennai, India.

First edition: October 2012

Contents

Figures

Acknowledgments

The following, more or less sequential accretions of influence, atmospherics, and administration impress and remain: My late mother Laura Glasberg, my father Dr Sidney Glasberg, Susan Gubar, Tom Roznowski, Jim McKelly, Tom Saya, Sharryn Kasmir, Robyn Wiegman, Don Pease, Eva Cherniavsky, Bill Mauer, Gabriel Walker, Vera Leier, Alistair MaKenzie, Jason Davis, Kathryn Yusoff, Christy Collis, Klaus Dodds, Laura Kay, Lisa Bloom, Connie Samaras, Anne Noble, Dawn Skorczewski, Adriana Craciun, Kerry Walk, Kristin Dombek, Marina Zurkow, Alan Nadel, Anna Bloom, Heidi Hoescht, Jasbir Puar, Alexandra Neel, Amit Rai, and Stephen Seely. My thanks to the series editors, Patricia Clough and Danielle Egan.

If I had been able to live up to either the runic intuitions or even the most simple directives of Taylor Black, this book would be better.

Guy Guthridge and the 2004–5 grant from the National Science Foundation's (NSF) Antarctic Artists and Writers Program carted and stowed my body for six weeks on ice. I will neither forget my time there, nor return.

Introduction: *On Ice*

And all the time you hear
The waves beat on that shore for a million years
go away go away go away
                    Katha Pollitt, "To an Antarctic Traveller" (1982)

Everything has changed around me since I first encountered the southern seas aboard Arthur Gordon Pym's hermaphrodite vessel, *Jane Guy*, in Edgar Allan Poe's 1838 *Narrative of Arthur Gordon Pym*. Poe's wayward drive into the "Hole at the Pole" theorized by nineteenth-century pseudoscience promised almost leeringly to "open to the eye of science" the mystery of the earth (*Pym* 201). Under Poe's perverse influence, my original plan to study US literary fantasies of the unknown end of the earth led me to ship's logs and maps as they were taken up by fiction writers in a feedback loop, as one French critic put it, a "journey to the bottom of the page."[1] The seafaring jacks-of-all-trades were replaced by professional heroes, navy navigators, government-appointed scientists, novelists, poets, and especially visual artists, whose work I was increasingly drawn to write about. Persistent fantasies of a hidden warm utopia of arable lands, mineral riches, and superior civilizations promising new myths of human genesis, intermixed with each report of icy, impassable seas.

The natural history sketches of the nineteenth century became documentary photography and later films in the twentieth century. Antarctica's cultural history might have remained a sort of antiquarian lever into national study, a gentlemanly pursuit of explorers of the "Heroic Age of Antarctic Exploration" (1895–1914) dominated by the 1912–13 "race to the South Pole" between Robert Scott of Britain and Roald Amundsen, remembered only so that they might be knocked off their plinths. After the Norwegian Amundsen arrived first at the South Pole and the British team died on the trek back to their base, the British seemed to own, if not Antarctica, certainly its value as a story of heroic failure and the end of empire (Spufford 1996). American explorers played no part in the pole drama; as one kindly but perplexed Americanist once put it to me, "No one is exactly waiting for the next book on Antarctica." And that prediction might have held if not for a phenomenon outside of the field of American Studies and the humanities: climate change and global environmental crisis.[2] Suddenly everyone knew at least something about Antarctica beyond the fact that it was cold: It was also melting. Southern ice now came with a ready-made politics.

We are in the midst of a new kind of Ice Age. The ubiquitous analogy "financial meltdown" sums up what Thomas Friedman's best-seller *Hot, Flat, Crowded* (2008) develops at length. Friedman's thesis is that the "parallels between what has been happening in the market and what is happening in Mother Nature are eerie" (25). More pointedly, Naomi Klein's "disaster capitalism" describes how "the politically unthinkable becomes the inevitable" through the creation and management of crises for which the distinction of climate forces from market forces seems increasingly impossible (Klein 2007). Even environmental activists appreciate—and await—"the power of a good catastrophe" to justify their agendas (Zengerie). And if the penguins are any indication, polar ice has become the site and substance of that most serviceable disaster. Documentaries like Luc Jacquet's Oscar-winning *March of the Penguins* (2005) and Werner Herzog's Oscar-nominated *Encounters at the End of the Earth* (2009), as well as animated features *Madagascar* (2005) and *Happy Feet* (2006) present these avatars of the continent as heroes and victims of climate change. "Disaster tourism" promotes seeing these adorably anthropomorphized birds before their icy habitat melts away.[3] However, rather than joining those anxious tourists in their southern cruises or the swooping aerial views of nature documentaries and their baritonal voiceovers lamenting the purity of the ice, or to follow in the melancholy footsteps of Antarctica's turn of the twentieth-century exploration dead, *Antarctica as Cultural Critique* reverses the flow of cultural production by beginning *with*—if not *in*—Antarctica.

Although the event of melting ice and its coordination with a narrative of climate change within a human timeframe has certainly captivated me, climate change is not the justification for this book. My long study of the symbolic lure of the unknown south has also too well alerted me to the ways that Antarctica as geographic terminus has been made to stand for hope or for doom—and usually both at the same time. Climate apocalypticism stakes much of its claims on data derived from research on Antarctic ice sheets, even as it holds up that same ice as a terrain of hope against the doom the data foretell. I appreciate as well the heightened notoriety of the 2012 100-year celebrations of the European attainment of the South Pole. However, the forces of the universe—elements, orbits, temperatures, and life itself—do not cohere around human observation and research time. The language and methods of mapping and measure of both the humanities and the sciences have too often unified in attempting to predict the unknown, or to manage a future that by definition cannot be predicted. The surveys, tracks, and narratives serve to defer a future, whether of utopia or doom, within the management of the present that itself may prevent the future it supposedly prepares. I have done my best not to allow a human perspective to become the default for Antarctica. Like Katha Pollitt's *Antarctic Traveller*, I heard Antarctica's icy shores endlessly repeat a simple message: "go away." It has been this provocation to desist that has produced this book on ice.

A Book on Ice

Ice is not to be written and not to be read. It is not to be captured within pages. It is not a book; certainly, it is not *like* a book. I say this because I have read many books on Antarctica whose titles more and more reflect the accumulating archive that is Antarctic ice. To list only a few: *The Ice: A Journey to Antarctica* (1986), *The Crystal Desert: Summers in Antarctica* (1986), *I May be Some Time: Ice and the English Imagination* (1996), *Time on Ice* (1998), *Innocents on the Ice* (1998), *Riddle of the Ice* (1998), *Pink Ice* (2002). 2011 has seen Paul Miller's *The Book of Ice* and Edward J. Larson's *An Empire of Ice*. A deep survey—or an ice core sample—of this accreting library of ice might still miss the substance of ice. Concerned with science, but not scientific in method, *Antarctica as Cultural Critique* is a book of ice that instead opens horizontally, "like a book open on its spine," which is how a glaciologist describes the striations of ice and earth in Figure 0.1. An ice book resists reading as detection of data and truth. It engages an open and living ecology—by which I mean a history of studying earth and its environments and their interconnections, as well as less conventionally understood forces, including the effect of the way these forces are studied. Such a book of ice is more of an assemblage of nested ecologies within the hard limits of the material earth.

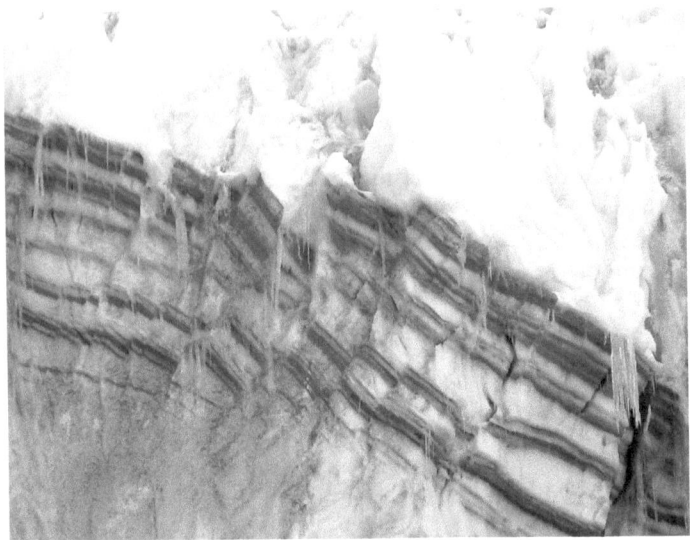

Figure 0.1   "Like a Book Opened on Its Spine." Striated glacier, Blood Falls, Antarctica. Photo by author.

In adding my volume to the impossible, shifting ice, I am thinking backward, from what Gilles Deleuze describes as a "perverse or depraved" and endless search "after signifiers":

> Or there's another way: you see the book as a little non-signifying machine...This second way of reading's intensive...There's nothing to explain, nothing to understand, nothing to interpret...this second way of reading...relates a book directly to what's Outside...Writing is one flow among others, with no special place in relation to others, that comes into relations of current, countercurrent, and eddy and other flows—flows of shit, sperm, words, action, eroticism, money, and so on. (*Negotiations* 7–9)

Deleuze is not the first to think of a book as within a flow of other books, of acts, and of life including nonhuman material life. All books flow into all books and acts, as water becomes ice and melts again and refreezes. Such circulations and strange openings—those Holes at the Poles—are what I hope to tap into, connect to, and proliferate. In writing about Antarctica, a place that has been overrun by heroic bodies and narratives, I want to retrieve ice from this history of loss at the "end" of modern time–space. I mean to indicate the possibilities foreclosed by historical overlay: the blankness or un-storied-ness that is not actually blank but rather its obverse, the inexpressible supplement to a historical process. History is inseparably, ontogenically different from becoming (Massumi 2005: 24). This book folds into the unforeclosed process of Antarctic becoming.

American Studies on Ice

Antarctica as Cultural Critique contributes to the valuable postcolonial scholarship on Antarctica by focusing on US and to some extent Latin American examples that extend and break with European models of coloniality. In the case of the United States, the failure of Anglo-American Antarctic exploration history to be read as a territorial extension of South America or of the southern hemisphere more generally demonstrates imperialism's ability to reconstruct space and time by connecting distant places and fragmenting formerly coherent territories, while it also points up the failure of imperialism to structure knowledge of the last continent into other than a Eurocentric map. Furthering a decentralizing of European power, *Antarctica as Cultural Critique* adds to scholarship on what new forms of biopolitical management look and feel like, particularly in nonnationalized territory and in territory outside of traditional sovereignty.

Geopolitical analysis cannot account for the ontology of ice; it can only enfold or make room for Antarctic space into an already-colonized globe by parcelling or sectoring out its ever-shifting nature. Postcolonial theories derived from native and anthropomorphic models create Antarctica as an exceptional, extreme, pure, outside territory. More accurately, this territory—like the under-theorized and visualized oceans—actually covers more of the earth's materiality than do its inhabited zones. *Antarctica as Cultural Critique* consciously avoids exceptionalizing Antarctica from the ecosystem of earth. Instead, this book examines conflicting modes of approaching, arriving at, and making sense of the ice, not as a symbol or as a new territory to be controlled, but as a material reality that itself has been part of how life on earth has been shaped.[4]

To arrive at the ice itself, it helps to consider how humans have been trying to live on ice. After World War I (WWI), the United States got back in the Antarctic game, after having sat out the Heroic Era. The major figure in US Antarctic exploration is Admiral Richard Evelyn Byrd. Although Byrd is known for being the first to fly over the South Pole, it is his establishment of a series of Navy bases on the Ross Ice Shelf that interests me here. "Little America" was a series of temporary colonizations; each iteration of the encampment had to be abandoned and rebuilt with the shifting ice.

Photographs of "Little America" reveal a world built of boxes and oil drums, all laboriously unladed from the supply ship. Rather than the earlier British-style wooden huts, the US Navy carved out an underground base more in the style of Amundsen, giving in to the only building material available—ice. Most remarkable, however, is the radio tower, an unprecedented feature of the US Antarctica (Figure 0.2). A metal structure that rose above the otherwise nearly indistinguishable profile of rooms carved out beneath the ice and scattered wooden boxes, oil drums, and equipment, the tower allowed direct communication with Washington, DC. Byrd's Antarctica was a futuristic amalgamation of sea, air, and disembodied technologies. While this technologic utopianism persists in Antarctica—a geodesic dome stood at the pole between 1972 and 2010—it also points to a past that is incomplete, and even impossible. That past and that incompleteness hinge on the intrusion of the human.

The problem of the human body on ice continues to be at the center of contemporary Antarctic geopolitics. Despite its patent unsuitability in the Antarctic environment, the human body is ever present in the landscape. The photographs of "Little America" demonstrate the complexities of human presence. They show men amid the things they brought to build a presence and to contain their presence. Little America was

Figure 0.2 "Radio Tower and Boxes." The Ohio State University, Byrd Polar Research Center Archival Program, Papers of Admiral Richard E. Byrd, image # 7802–2.

an extreme colony, unlike a military encampment and unlike previous European environments. Byrd had hopes of an official claim based on his explorations and encampments, going so far as to overwinter alone in a hut in 1938, almost dying and necessitating a rescue. Despite Byrd's campaign, a succession of US administrations declined to produce a legal territorial claim. Consequently, Little America, what historian Paul Carter called a "town at the end of the world," both extended and undermined notions of territory and nation.

The photographs of Byrd's expeditions that appeared in a variety of popular magazines beginning in the 1930s and throughout the 1950s emphasized US national symbolics without any legal substance. The very iconic photograph of navy officers with a US flag presents the sort of flexible and incoherent national presence I am talking about (Figure 0.3). Aside from the tower, a flag pole was the only other feature rising above the ice. A flag pole was a difficult symbol to maintain amid wind and the cold and unpredictable sea ice. In this photograph, without an explanatory caption, it is impossible to know if the flag is being raised—or lowered; if this is the end of the earth or the beginning of a new phase of US empire. What is even more interesting is the state of the flag itself. Like a flag in battle, this

Figure 0.3 Flag at "Little America," 1928. The Ohio State University, Byrd Polar Research Center Archival Program, Papers of Admiral Richard E. Byrd, image# 233–2-130.

star-spangled banner bears the imprint of hard usage and struggle. The symbol of nation is not whole; its very tatters are meant, however, to resuture US nation in this unlikely outpost. My question in this book is more focused on the tatters, the holes, and the incompleteness of the flag. Byrd's tattered flag and his beaten body are the new symbols for an Antarctica that is falling to ruin, even as its historical associations have been as epistemological completion of the globe, internationalism, and the utopian beyond. By pursuing the holes in the flag, this book produces a possibility inherent to ice and outside the symbolic suturings of nation.

In working against powerful national conceptualizations of ice as inimical or frozen or deadly or as a blank to write on, *Antarctica as Cultural Critique* also resists globalist conceptualizations of Antarctic ice as either a laboratory or repository of data, or as a pure wilderness. Thinking of Antarctica instead through the materiality of ice offers ecology as an alternate mode of connection. My use of the term ecological departs from the implications of environmentalism as the interdisciplinary management of earth's environment. Rather, ecological suggests the nested and looped flow of interconnection that includes a modern environmental awareness.[5] The emergence of Antarctica and its connection to climate disaster both through its position as the end of the earth and its geophysical makeup as ice has upset the drive for convergence and connection once offered by imperial teleologies. This crisis of climate mobilizes governmental control in which capital thrives, even at the expense of the populations in whose name it serves. "Disaster capitalism"—even in Antarctica's seemingly free expanses— has been an incitement to remilitarize through scientific regimes of surveillance and the production of a new population of science workers. Hope and futurity, notions so often tied to Antarctica's relatively unencumbered ice, become aspects of a neoliberal economic and corporate– military power strategy.

The motive behind *Antarctica as Cultural Critique* ironically coincides with the pronouncements of (Ret.) Admiral Conrad Lautenbacher, an employee of Antarctic Research Services, one of several transnational corporations that in 2011 vied for the National Science Foundation (NSF) Antarctic support contract. Lautenbacher avers that "people need to learn more about what's going on down there." *Antarctica as Cultural Critique* departs from scientific internationalism's bland promotion of knowledge, arguing most broadly that Antarctica's incorporation into the globe is a visually mediated violence taking the form of a series of blank outs. As the blank space of the imperial map, as pure wilderness, malevolent whiteout, or as a dream of untapped resources,

Antarctica ensures both the future and the end of the mapping of territorial empire. Such blanked-out promise reproduces time as possibility and makes space virtual. Precisely because of this imbalance between what citizens know about Antarctica and how much money is invested there, now is the time to introduce Antarctica into the conversation on US empire.

The Most Mediated Place on Earth

More than for any other place on earth, visual mediation defines and has created the territory of Antarctica. Indeed, being on ice has for the most part meant seeing ice, and producing and circulating visualizations of ice (Glasberg 2005). For the explorers of the Heroic Age, photography helped establish proof of their claims of arrival at the South Pole, thus proliferating visual imagery of a place experienced by so few on behalf of the mass of humanity. Due in part to the legacy of human-centered travail and hardship, bodily difficulties that extend to optics in prevalent and even defining Antarctic discourses like whiteout, the contemporary problematic regarding Antarctica resides in the tension between Antarctica's hypermediation and its historical and material challenges to human sight and understanding. Antarctica is thus the most-mediated, manipulated, surveilled territory on the planet. Yet, because it has been experienced by the smallest percentage of humans, it is commonly thought to be the least known. More accurately, Antarctica is a surveilled and controlled territory that is a species of nonterritorial knowledge possession of a succession of expert actors, from the national heroes and military men of the twentieth century, to the adventurers, scientists and support workers, tourists, and state-supported artists of the present day.[6]

Antarctica as Cultural Critique relies heavily on—in fact, insists on—the mediation of art, especially visual art. Engaging the photographic archive and the visual art of Antarctica opens an alternative indexicality, or way of reading the trace of the body, on ice. It is a record that is particularly open to a "new materialist" approach that opens matter to flows and processes not reducible to an outside of the physical (Coole and Frost 2010). Such an approach also engages with ice in ways not necessarily mediated through presence, science management, tourism, or other conventional modes of seeing ice or being on ice. I attend to not only how to see Antarctica as a polar region and as a territory incompletely corralled by a gendered human-centered vision. More crucially, my interest is in destabilizing the visual record through the haptic, embodied sensorium that has left fossil traces in photographs, on bodies, and on the ice. The tension between seeing and being in Antarctica and between forms of mediation and a specifically European masculinist history of presence broadly characterizes this book's approach.

The Antarctic Post-Heroic

Signs of a greater awareness and interest in the territory are everywhere in the news, in popular films, and television. Danger, risk, and sentimentalized notions of testing human endurance have been the lingua franca of Antarctic endeavor, carryovers from Europe's colonial drive. These notions of struggle and survival literally mark the inhabited ice with crosses symbolizing the passings and achievements of humans. The language of achievement has itself passed from imperial registers to those of science. Not only has this passage been seamless, it is ongoing. Scott, the self-sacrificial hero of British empire (who has also been in turn a sacrificial object of postcolonial critique), continues to call forth accusers and defenders. Recent books reconstitute Scott and his expedition culturally and in terms of science. David Crane's *Scott of the Antarctic: A Life of Courage and Tragedy* (2006) rescued Scott the man and icon of British empire from his postcolonial and feminist detractors. Edward J. Larson's *An Empire of Ice: Scott, Shackleton, and the Heroic Age of Antarctic Science* follows Susan Solomon's *The Coldest March* to rehabilitate Scott as the cultural icon and explorer through science. Solomon, a major climate scientist and codiscoverer of the ozone hole over Antarctica, identifies in Scott a worthy precursor unappreciated for his scientific rigor amid what Solomon demonstrates with atmospheric research were unusually harsh climate conditions. Larson's title reclaims the tarnished term of empire by rerouting it through science. His is the major recuperation of the Scott legacy in the name of international science. This phase of the cleansing of the heroes of empire implicitly relies on the growing concern for climate change, a concern that also justifies international science and much of the human presence on Antarctica.

Encounters at the End of the Earth, a documentary by Werner Herzog made with the support of the US-National Science Foundation (NSF), which was nominated for an academy award in 2009, demonstrates some of the complexities of what I am calling the post-Heroic. *Encounters* is framed with footage of the underwater realm beneath the ice. Herzog enchants the under-ice footage (Herzog did not directly experience this view) imposing a soundtrack of chanting church music. Interviews of workers and scientists centered at the US McMurdo base anchor and provide foils for Herzog's personal dramas in attempting to find a place for himself (and by extension, all humanity) in Antarctica. He makes no bones with manipulating images, cutting off conversations, roughly framing interview subjects, using sci-fi animation and allusions—his entire bag of tricks—to impose his vision.

This vision is of a human history on the planet coming to an evolutionary and environmental crisis. Herzog's anger about humanity's coming catastrophe bleeds into his own frustrations with the layers of representational and logistical apparatus between him and a direct relation to Antarctica. Herzog was wishing for something more exciting, something out of Poe—explosions, starvation, lost tribes, or at the very least the good fortune of a timely explosion of lava from live Mount Erebus. Instead, what he got was the most surveilled, safe, and documented journey he had ever been on. Antarctica is indeed a difficult landscape in which to navigate and Herzog found himself necessarily within the control of Raytheon Corporation and other NSF logistics subcontractors. Finding that he is no Shackleton or Amundsen, Herzog laments the heroic past and mocks it too, incorporating Heroic Age promotional footage of men in expedition gear that had been (ironically) shot on a sound stage in England. Herzog mocks and yet longs for an impossible masculine heroic ideal, which he implies the "original" heroes also could not achieve. However, in the end, the filmmaker's imprint prevailed—Herzog found his disaster film. Modern Antarctica is feminized, not by traditionally sexed bodies or even by gender. Rather, it is technology and science itself by coming after the (supposed) direct encounter with the ice of the Heroic Age that has undone Antarctica's possibilities as a wilderness or ground for reproving human survival. Herzog, luxuriously and safely disgusted by modernity, stages danger and eruption through editing and by leading his gullible subjects into self-serving heroic and unwittingly gruesome or silly accounts of the environment or their activities.

The highlight of Herzog's search for Antarctica's dangerous spirit comes in a scene on Erebus, a site routinely visited by helicopters from the nearby US base at McMurdo Sound. Ostensibly interviewing an enthusiastic volcanologist, Herzog allows his camera focus to drift, distracted. It lands on a remote surveillance camera. Camera captures camera in this face-off of technologies of visualization. If the hero is dead, stifled, or worse, ridiculed, Herzog offers a new hero for a new Antarctica: the camera, the disembodied seeing of the same camera model it turns out used for prisons. The model-10099, boast the scientists who use it to see for themselves, can survive a prison riot. In this riot-proof surveillance camera standing guard over the volcano, Herzog finds an ironic connection to quasi-militarism and a new hero of scientific management, one that may very well outlast even the scientific establishment. Herzog cleverly plays off the failure and the promise of the post-Heroic: Humans will perish outright of their own ecological mismanagement, just as surely as they will be outlasted by the products and modes of their technological accomplishment that fed the disaster.

While Herzog camps up his role as Antarctica's new environmental hero, most commentators play Antarctica's role as a canary in the mineshaft of ecological disaster straight. The sentimental futurity often produced in the name of Antarctica's uninhabited expanses comes under pressure in *Antarctica as Cultural Critique*. Antarctica can be used, felt, and organized in multivalent ways when it both promises the future and threatens us with it. While recent scholarship somewhat moves beyond exceptionalizing Antarctica's melancholic masculinist heroic exploration, colonial analysis' emphasis on human arrivals and inhabitation itself contributes to the fantasy of Antarctica as a place "frozen in time" and thus endlessly available for an instrumentalized future. Antarctica's historical availability to projections and other temporal lag times has opened it to connective flows with virtual territories. The heroes are still arriving. But today, they do not leave the kind of footprints that once indexed heroic geography. Virtual arrivals, affective assemblages, and massively distributed weather effects connected to global climate change continue to remake Antarctic space. What I term the "post-Heroic," theorizes an affective relation of resentment and desire that forecloses even its own resistance, resulting in restagings of a reviled imperial past within a neoliberal present that is supposedly cleansed of those associations.[7]

Antarctica, though preserved from legal claims until 2049, is in fact increasingly occupied by national science programs and other nonnational entities. The status of humanity on Antarctic ice is at once highly assumed and under-theorized. That is, legal and scientific regimes purport to control and manage access to Antarctica in the name of environmentalism and global equity for non-first-world nations, but have absolutely no commitment for keeping Antarctica free of harmful development from humans. In fact, the very existence of regimes such as the Antarctic Treaty System (ATS) and agencies such as the United States Antarctic Program under the NSF mitigates against curtailment of development on and in Antarctica. This condition of managed and increasing human activity in Antarctica constitutes another level of what I argue is Antarctica's extreme mediation. As a shared territory under an international treaty inhabited by an infinitesimal percentage of humanity, Antarctica's becoming took place in the name of populations, but not through them. Today,

Antarctica is a place without people that is used in the biopolitical management of populations—in the rest of the world—through survival and futurity and through affects of belonging to community and to territory. I contend that because the "bio" in Antarctica's biopolitical management is distinctly non-anthropomorphic, it cannot point to routine humanist and gendered myths of origin and belonging. Yet, its exploration history spawns (with few exceptions) endless iterations of the edict to survive to justify an array of state and international coercions.

On the one hand, humanity projects onto Antarctica its desire for pure origins, fantasies intensified through commonsense (if erroneous) associations of the materiality of ice as visually white and with its preservative properties. Such northern-centric misconceptions of and with Antarctic ice have fed perhaps the most troubling and often repeated cliché: that Antarctica has never known war. While such an observation, given the continent's lack of natives, is meaningless, the patent absurdity of the statement requires only the example of the 1982 Falklands–Malvinas War between Britain and Argentina. The effect of such statements about Antarctica's peace and purity—what might be called its exceptional status—requires a more complex discussion. Despite treaties and the grand enlightenment of modern times, the polar regions are still proto-war zones and the ice remains a perfect backdrop, a sublime incitement for a heroic masculinity and militarism that may endure longer than the very icy wastes an earlier generation of British heroes failed to dispel. The post-Heroic operates as a reinvigoration through repetition and lament of the seemingly less-compromised colonial regimes of exploration. Understood through the screen of the post-Heroic, the timing of Antarctica's exploration takes on layers of meaning. The South Pole may have been the last place on earth, but its achievement in 1912, far from completing an imperial project of globalization, actually signalled the dead end of the map of empire long before its official breakup in the 1960s. Thus, the end of the terrestrial earth at the South Pole serves as a temporal and geopolitical repositioning of postcoloniality—a proposition that this book proliferates.

As the masculine body, along with its ethical and legal certitudes, lost its centrality to this endeavour, a complex series of critiques, replacements, and recuperations around that loss have at various times revived the melancholic human claim to Antarctica. The desperate and failed management strategies and approaches to ice described in *Antarctica as Cultural Critique* are both products of and responses to the paradoxical nature of the drawn-out crisis of geopolitical power and the environment that has sadly yet finally given value to the ice.

Chapter 1, "Antarctic Convergence," heads straight for the obvious. Most books on Antarctica begin with a history of its mapping and predictably arrive at the present-day Antarctica—and of course, the book unfolding (or scrolling) before the reader. However, this kind of literary mapping entrains and assumes processes that really need to be questioned and that indeed may not even apply to a territory of ice. If there is anything to be ultimately gained through a "survey" of Antarctic mapping, it is that no mapping is commensurate with the territory and that all mapping constructs a particular territory for the uses of the times and the cartographer (or writer). Mapping Antarctica opens up new territories as it forecloses others, a process that continues into the present era.

"Google Earth Antarctica" uses geographic information system (GIS) mapping to promise a cutting-edge map that connects Antarctica smoothly to the rest of the earth. Strangely, the greater accuracy promised by GIS exaggerates the problem of Antarctica's historical and geographic position as the end of the earth—or its problematic placement as supplement to a total globe—by creating a new "blank space" on the map. Responding to the metageographical force of prolepsis and deferral at work, practitioners of "experimental geography" such as Peter Fend and Trevor Paglan use satellite mapping techniques as well as data as counter-mapping to reveal the types of dispossession ongoing in the governmental control of territories for which population—national or popular—has no set relation.

Building on the ways gender-neutral humanism distorts what would seem to be the most populist mapping of Antarctica, chapter 2, "Refusing History: After Ursula K. Le Guin's 'Sur,'" argues that even Antarctica's feminist and postcolonial revisionist exploration history has been overrun by positivist, evidence-based scholarship that works to recuperate the racial, gendered, and national hierarchies it purports to redress. Le Guin's 1981 short story opens up layers of possibility within the bare colonialist history of exploration by inserting South American women discoverers at the South Pole, who arrive before the men of history but decide never to let their achievement be known. Le Guin's description of the South Pole as a "bindu" discomposes Antarctica's masculinist Euro-centric histories with more wit and economy than have volumes of postcolonial criticism ("Heroes"). Le Guin's 1970s-era ecofeminism holds up so well because it is anchored to a speculative respect for the radical otherness of Antarctic ice within the still unknown ecologies of this planet.

Chapter 3, "Who Goes There?" traces three fictional versions of the discovery of an alien buried in Antarctic ice through pre-WWII, post-war, and the contemporary era of scientific management under the ATS. Byrd's career-long struggles to promote an official US claim demonstrate the nature of the post-Heroic in Antarctica. The post-Heroic describes the horrifying desire for a heroic status that no longer can be achieved in an era of international science, demilitarization, and the breakup of homosocial work groups—changes that took place more or less following postcolonial and other legal liberalizations throughout the world. The myths of Antarctica's heroic past continue to hold sway over workers in contemporary Antarctica, who are unable to achieve the status formerly accorded Antarctic exploration and so find themselves caught in a feedback loop of management and surveillance that they rationalize through affective attachments that attempt to recoup a sense of heroic individuality once symbolized by Antarctic endeavor. The speculative horror of Antarctic fiction both covers for and indicates the more mundane, yet inexpressably disturbing facticity of neoimperial control over a territory that is nominally and in the imagination of the public and of those working on ice, a continent free of international strife, national occupation, and economic development for corporate profit.

How transnational corporations sidestepping the protocols of the ATS—and perhaps even more significantly—profit from Antarctic ice is the problem analyzed in chapter 4, "On the Road with Chrysler: Virtual Capitalism and Empire without Territory." The strategies of a series of national advertising campaigns featuring Antarctica by transcorporations from Chrysler to IBM to KBR, Inc., I argue, produce Antarctica as a virtual location for the negotiation of transcorporate power and profit by rescuing oil capital from its own limits. The advertisements aestheticize Antarctica's wilderness and depend on maintaining a fiction of Antarctica's extremity and even exception from the ravages of expansion and development they in actuality represent. Under the ATS regime of nonclaimancy, Antarctic empire has become more thoroughly indexical as a trace, a physical memory standing in for the lost object of possession, and even for some nostalgic object of empire purified of its ruined encounter with colonialism. But indexical empire is not reliant on the coordinates of the human body as a referent. A new type of indexicality that uses movement and process rather than fixed or permanent location now has developed. From the respatializing of the continent produced under the ATS emerges the re-territorializing of the new road to pole. This new supply-chain-type road solves the problem posed by article IV of the Treaty deferring claims: The road to pole—its management and constant upkeep as well as its imagined trajectory of movement—replaces the territory itself. This claim without territory can be understood as in part motivated by rising fuel costs and the anxieties of the end of oil capitalism. The corporate–logistical road retains the trace of Byrd's masculine heroic body struggling to survive on ice, while also exploiting environmental crisis to become a form of corporate possession whose supply chain also includes the production of an idea of the future itself.

Chapter 5, "Photography on Ice," follows the tracks of capital's vision of Antarctica as a space of primitive accumulation, or as the permanent outside of the capital system—a way of seeing Antarctica that converges visual, market, and cultural economies. This connects to the earlier discussion of disaster capitalism and territorialization without territory as it converges on Antarctica as an anxious site of climate change—and as an opportunity for profit. Relevant to my analysis are powerful formations of Antarctica as a blank page, *tabula rasa*, white space on the map, or an empty screen for the projections of culture and of "blanking," to use Naomi Klein's description of how "disaster capitalism" searches out and creates spaces of primitive accumulation that are also forms of dispossession.

These visions of instrumentalized or incorporated Antarctic landscape use similar techniques as seen in the aesthetically normalizing photographs of the American conservationist Eliot Porter, one of the first artists in Antarctica supported by the NSF in the early 1970s. Even more importantly, the production of Antarctica as an exceptional place that coheres a range of global concern comes across as less benign in NSF-supported photographer An-My Lê's landscapes of an industrialized South Pole marked by traces of militarization and discordant material and affective traces of suburban environments. Antarctica's alignment with neoliberal militarism is emphasized by Lê's inclusion of the US bases at McMurdo Sound and the South Pole in her 2009 series, "Events Ashore" tracking US military bases worldwide. Antarctica's connection to militarism and industry also comes through repertoires of sci-fi and fantasy. A reading of Robert Smithson's little-known photo collage, "Proposal for a Monument in Antarctica" (1969), depicting a massive industrial ship anchored on an icy shore, ends the chapter. In the era of the Apollo moonwalk, Smithson's melancholic monument to mechanical reproduction and to fordist production looked both backward and to a future in which neither human presence nor traditional notions of earth anchored the "earthworks" he was to make famous.

Chapter 6, "Sculpting in Ice," considers a new form of environmental protest art focused on climate change and on its quintessential marker, ice. Through photography, science fiction narratives, and computer-generated animation, the feminist artists discussed—Anne Noble, Connie Samaras, and Marina Zurkow—challenge aesthetic, national, legal, and global scientific regimes that both define and constitute the controversy over climate. I propose the term "anthropogenic landscape" to describe the fields of vision produced within a nexus of competing forces: the material earth; the built environment and its products; laws of physics and optics; human-centered technologies of vision and reproduction such as photography and mechanical and computer animation; and knowledge production, in the lab and on site. An anthropogenic landscape reflects consequences and processes driven by human agency and human-generated sight. Yet, it may circulate possibilities beyond the human and seek to be formed by or even to produce nonhuman perspectives and futures.

The anthropogenic landscape produced by these artists falls within a tradition of feminist protest art that is now taking up climate change and data as violences of mediation. These mediating violences include active governmental suppression of images, data, as well as more passive forms of deterrence and avoidance in the production, analysis, and circulation of information on climate change. Specifically, these artists incorporate into their own practices and products the problem of *how* ice becomes data, responding to reports of climate change and ice melt, and in some cases, producing data through art practice to challenge the suppression or even the basic lack of evidence of climate change. Samaras and Noble use photography to retrace the human imprint on the ice through capturing light in ways that suggest the co-construction of human and nonhuman (or alien) perspectives on changes in the environment. Zurkow's digital animation loop replaces the degraded ice with a virtual polar sea whose endless lure only begs the question of the decay of even the virtual. All these artists, in questioning androcentric and science-driven knowledge formations, open toward an ontology of ice from which a new form of social protest media may emerge in the circulation of images of polar landscape. In encountering traces of the human in the land, Antarctic landscapes do more than question an optical–rational perspective. They also encounter—if not produce—the limits of technology, of vision, and even of seeing itself.

The epilogue, "Becoming Polar," begins with the end of a commodified and abject Antarctica. Before the modern era, the portion of the mapped globe ascribed to Antarctica shrunk as geographical information of the region accumulated. Antarctica became a strange supplement to ongoing colonial and economic activities. In the present phase of globalization and convergence of climate and disaster, Antarctica is again shrinking as data on its melting ice accumulate. However, more than shrinking, Antarctica's ice is redistributing; it is on the move. As ice melts—or migrates—it becomes less itself, changes form as water gradually inundates the globe. Melting ice along with the hard limits of oil capitalism create new conditions of precarity all around the globe. Desertification and extreme oil drilling once were features of polar environments, but in the desperation for profit under the belief that oil capitalism can be maintained, corporations seek profits and court disasters that once were unthinkable. Hope as futurity—and doom as only another version of futurity—converges and disaggregates in Antarctica as data, supply chains of people, narratives, machines, and beliefs flowing through what Le Guin named the "living ice."

Chapter 1

Antarctic Convergence: The Problem of Antarctic Mapping

Mapping is the cultural history of Antarctica. Before the ninteenth century, mapping alone constituted the biopolitical relation to the territory. Due to the time lag between a southern landmass being imagined and its actualization through exploration and discovery, mapping has had a very specific, even disproportionate importance for Antarctica. Declared in 1895 as the last undiscovered place on earth by the Royal Geographical Society, and lacking natives, Antarctica has been preeminently defined by its mapping. This chapter concerns the distorting effects of modern global mapping, from the development of the post–World War II Treaty System's sector mapping of the continent, to the geographic information system (GIS) mapping of Google Earth Antarctica, a citizen-participant, open-ended satellite–terrestrial mapping.

Antarctica's mapping has been about more than simply knowing what we now designate as Antarctica on maps. For each discovery or notation of Antarctic geography also caused a change in the way the globe itself was known and mapped. At first, these changes were profound: Antarctica's virtual conceptualization by the Greeks as the southern land helped produce earth as a globe. Throughout the centuries navigators charted—or more importantly did not chart—land that gradually extended the not-Antarctic as much as it shed light on the nature of the putative region. As the great southern continent shrunk, the rest of the globe grew until finally men walked to the South Pole and the mapping of earth seemed complete. Mapping Antarctica has thereafter been not only a matter of filling in knowledge, but of signaling a relation to the rest of the world and among the territories of the rest of the world. Antarctica since the 1960s has figured as a site of convergence, or for a "coming together" of the entire globe.

Historian Edwin Mickleburgh writing in 1987 exemplifies a growing ecological imaginary attaching to a completely known Antarctica:

> Surviving in such extreme conditions harboured the beginnings of a new approach, changing the way man saw himself within the intricate structures that determine the natural environment and thus his own well-being. There is a glimpse of a possible re-orientation of the values that has its genesis in Antarctica, a way in which man might come to regard earth as a whole, politically, economically, and environmentally. (*Beyond the Frozen Sea* 1987: 5)

Mickleburgh is remarkable in his linkage of Antarctic geography and the environmental "extreme conditions," through the "well being" of the colonizing human body, to a sense of the earth as a "whole, politically, economically, and environmentally"—or what we now call globalism. Mickleburgh offers a modernist political/environmentalist view of the significance of Antarctica as a "symbol of our time" (15) and as a material place in need of protection by the concerted efforts of the rest of the world, which its discovery marks as complete.

Antarctica is more than ever this symbol. But the times as much as the symbol keep shifting with the ice. Other scholars began to emphasize the process through which this completion occurred. Antarctica's preeminent geographer, Klaus Dodds, takes a retrospective look at the process of Antarctica's mapping, warning: "In the era of remote sensing and satellite photography it is easy to forget that knowledge of the Antarctic was fragmentary in the politically charged post-war period" (*The Geopolitics of Antarctica* 1997: 223). Dodds emphasizes the problem of integration of the technical modes of Antarctic knowledge. Together, these quotes present Antarctica as a paradox: the congealing of ice or convergence of disintegrated parts. The "fragmentary" and unintegrated, as Dodds suggests, have been ideologically buried by their forgetting, or their disorganization from memory. Dodds is correct to point out the messy and the fragmentary processes of creating territory as integral. These processes, as the phrase "easy to forget" suggests, are connected to memorialization whether through cultural history, the built environment of Antarctica, or the bodies of the people moving across or living on ice.

Modernist tenets, of the integrity and wholeness of the earth and of knowledge and of technology as the agent of this whole truth's unveiling, persist in contemporary knowledge production around Antarctica.[1] The fragmentary, partial, lost, unrecoverable, injured, or mysterious have been forgotten in favor of narratives of Antarctica as revealed, knowable in itself and as part of the world system. International accord is the rule of the day, with science's interdisciplinary and testing modalities as the engine.[2] For Mickleburgh and others writing in the wake of the nongovernmental group Greenpeace's invigoration of ecological awareness in the 1980s through and on behalf of Antarctic ice and waters, Antarctica is the key to the geopolitical and economic world system. Antarctica has become an exceptional place—meaning its distinction from the other six continents has beome an alibi for an array of geopolitical repressions—whose rescue from the fate of the rest of the planet becomes talismanic of global survival.

But a globalist, ecological imagination sacrifices aspects of the very world systems it seeks to protect; it cannot see the tatters of the flag as the actual territory. As discussed in the Introduction to this book, systems such as nation repair the flag even in photographing it, even in the very performances of raising and lowering it. But this tattered flag—like the incomplete map and the battered bodies and minds of the early heroes such as Byrd and Scott—is and remains Antarctica. Climate crisis today is an actualized threat that in its massive distribution though time and especially the space of maps and bodies and memory is hiding in plain sight. Integrationist, management views suture over the persistent, material, symbolic, and epistemological destabilizations posed by the last place on earth to the very modernist epistemologies that have given Antarctica its teleological weight.

Yet, amidst the ever-growing talk of globalism and the new interconnectedness of nations, markets, and cultures, as well as the admiration for the multiple possibilities of permeable borders, there remains a rather familiar notion of what constitutes a place. Place is understood in relation to populations and their specific cultures. On most maps, places without people register little more than blankness. Antarctica on the typical world map remains a distorted elongation lining the bottom of the world. With no indigenous population, no human history, no accessible resources, and no landscape to be apprehended under aesthetic visual regimes,[3] Antarctica fails the test of place. How then does one begin to think of this last-discovered and least-known continent as a proper subject of globalization?

From Roald Amundsen of Norway and Sir Robert F. Scott of Britain, we learn how discovery can actually mean the end of knowledge, not its triumphant beginning, and in this, we are driven into the heart of modernity's own anxieties about epistemological security.[4]

While Amundsen's and Scott's race to the Pole in 1911–12 roughly coincides with some of the most prolific and Utopian years of aesthetic modernism, it is significant that as the aesthetic realm becomes fascinated with surrealist and antirealist representational modes, the technological advances wrought by scientific and military exploration increasingly emphasize visual knowledge based on replications and enhancements of the human eye.[5] In this regard, much of the current critical attention to the "posting" of the human body in representational practices—the rise, that is, of the "virtual" as an arena of commerce and production—might be traced to certain technological changes wrought by the twentieth-century development of military and scientific exploration.[6]

In this history, Antarctica is that place outside the circuits of the known world that both precedes the moon as a destination of otherworldly knowledge and is coterminous with "outer space." Once the globe is figured in its totality, in what is now a commonplace of mass-mediated imagery (an isolated sphere hung in celestial space), one way to manage modernity's gesture of arrival and progress is to imagine the place beyond, to gaze back at an earth left behind. Antarctica as the last place on earth is also, then, the first place for the technological development of new representational practices, new modes of exploratory knowledge, and new ways of tying the unusable (or wasted) earth to the engines of capital accumulation that have propelled us into transnational global exchange.

Antarctica has never been an easy place for humanity to attach to preexisting flows of culture. Chief among its historical discontinuities, Antarctica is not now and has never been a national space. While the globe was coming under the reign of sovereigns, and later of nation-states, Antarctica was remote as myth until the turn of the twentieth century. By the time of Antarctica's belated attainment in 1911, nationalism was so widespread and yet an unstable mode of geopolitical interaction, that its pattern resembled the crazed ice fields of the southern oceans. We might even say that the pursuit of the South Pole by the West indexed the ideological conflicts within the field of the imperial nation, and further, that its discovery inaugurated the search for posthuman technologies of empire-building that would dominate the globe by the century's end. The pursuit's climax, the race to the South Pole between Britain and Norway, became a sort of overdetermined performance of a national symbolic, a conflicted attempt to bolster the narrative of progress implicit in empire.

In 1911–12, people around the world followed the race between Amundsen and Scott to plant the flag of their nation at the crosshatch of latitude and longitude that was the South Pole. The drama of that race remains riveting to this day. Amundsen, with his penchant for wearing bowler hats (when not imitating the polar survival methods of the Inuit), ran his expedition as a model of order and functionalism, feeding sled dogs needed for the trip to the pole to the remaining dogs for the trip back. Scott has gone down in history as a romantic figure, a throwback to a more lush age of terrestrial opportunities in which explorers were poets and theologians as well as determined sled pullers. Scott's notion of British glory and empire could not be achieved on the carcasses of poor sled dogs; so he and his men hauled their own sleds to the pole, where they discovered the Norwegian flag Amundsen had planted several weeks earlier on his arrival. Perhaps, it is Scott's poetic death, freezing on the polar plateau with a pen in his hand, that explains why Amundsen today is strangely a footnote in history next to Scott, who continues to be the subject of memoirs, histories, and fiction. Taken together, we can read Scott's failure and Amundsen's triumph as emblems of the impossibility of discovering Antarctica within the register of imperial exploration. There is no picture to send back from the pole except the endless, anti-landscape "white out" that defines the continent as a geographical space; there are no witnesses beyond the explorers themselves. The idea of Antarctic discovery was grand enough, but the material facts could not bear the weight of the West's long anticipated arrival.

Despite the limitations of the national in the Antarctic problematic, Norway and Britain continue to base their Antarctic claims on the promissory represented by the explorer's flag propped up by chunks of ice on an Antarctic plateau. While the United States has never officially entered a claim, Argentina, Chile, and Australia have each recorded Antarctic claims of varying degrees of historical, geographical, and rhetorical persuasion. International law recognizes territorial claims based on discovery, conquest, or settlement. Since Antarctica lacks an indigenous population either to make claims or to make conquest upon, claims of originary or early discovery have been marshaled instead. Britain rests its claim on Captain James Cook's 1772 expedition. The United States, similarly, trots out Nathaniel Palmer's 1823 sighting and Captain Charles Wilkes's 1841 landing. Norway naturally glories in Amundsen's first attainment of an inland goal. And of course, the industrialized nations have been erecting stations and deploying military and scientific monitors all over the continent. Countries with less-developed imperial histories, such as Argentina and Chile, base their claims on geographical nearness, and in Argentina's case, on colonization by civilians, which in 1979 produced the first human born in Antarctic territory.

But even by the late 1940s and early 1950s, with territorial claims amassing, an international agreement emerged, if not to supplant the logic of nations, then to manage their competing claims. Developed during the Cold War as part of the 1959 International Geophysical Year, and re-ratified in 1993 under the latest national reconfigurations, the Antarctic Treaty System (ATS) bans military, nuclear, and industrial development in the continent.[7] Essentially, the ATS defers competing national claims into the future while administrating Antarctica as an international scientific zone. The ATS functions as a way of suspending national claims while deferring the very logics of territory as national entities while simultaneously bolstering those national agencies that seek to govern and control the possible material wealth of the continent. In this sense, the ATS holds Antarctica in suspension for a future nationalism. Yet, all attempts to "rule" the area in terms of nations have been incommensurate to the territory. Given its lack of indigenous population, its relative remoteness in time and space, and the fragility of its environment, Antarctica throughout the twentieth century has thwarted the ideologies, law regimes, and operating practices of nationalism. Most crucially, the ATS today is out of sync with both ideological and economic modes of production that accompany contemporary capitalist accumulation. Tied in some sense too tightly to a modernist understanding of territorial bodies, the ATS asks Antarctica to stand still as various transnational corporations develop new representational means to make of the seventh continent a valuable resource.

Before we arrive fully at the present, however, a few words are necessary about the symbolic and material interaction among Antarctica, knowledge, ideologies of nation, and modes of production that accompany Antarctica in the global consciousness today. By the time the West revisited Antarctica after WWI, the Western world had been transformed into competing blocs of multinational economic power. The economies of the emergent first world were more connected and interdependent than ever before. In this context, the vast land of Antarctica was once again sighted (discovered anew), not for colonization in the traditional form, but as a strategic site for national power in the quest for world economic and military dominance. The era of bipedal exploration technology quickly seemed quaint as the footprint gave way to aerial surveillance. The human-based forms of knowledge that were linked to empire—colonization and its scholarly ally, anthropology—would be displaced by posthuman globalizing strategies and the disembodied surveillance made possible within twentieth-century transnational capitalism.

The major figure in the shift away from human-centered Antarctic knowledge was an American, Rear Admiral Richard Byrd, whose aerial mapping of the continent extended the scope of the human eye beyond the limits of the terrestrially situated body. It was this very notion of the cotermineity of human eye and a more extensive "god's eye" view that allowed the shift from human-centered and powered technologies into the robotic surveillance devices—ranging from early-twentieth-century weather balloons to Galileo, the robot sent in 1994 to explore Antarctica's active Erebus volcano—that depart in significant ways from the modes of human-based information. The difficulties of imposing human-based technologies of seeing, sensing, and knowing highlight the ways the materiality of Antarctica has shifted with and under technological changes and epistemological arrival. By the twentieth century Antarctica, the long-deferred completion of the globe, has itself become global.

A Metageography of Globes—Convergence as Metaphor and Map

Because Antarctica holds a position at the southern pole of the globe, its history of mapping has interlarded with the emergence of beliefs about the globe. Understanding the metageography of globes or the cultural assumptions of their mapping is crucial for the process of Antarctic mapping. Mapped globes offer a fantasy of completion. They are by definition representations of an entire earth. Yet, they escape neither partialness nor partiality. As geographer John Agnew has argued about globes, "In masking their selectivity behind empiricist claims to accurate representation they provided a powerful means of picturing the world as a whole as if it existed independently or separately of all attempts to conquer, tame, and exploit it" (19). Thus, globes tend to obscure their own partiality and their political usefulness.

The modern globe was a "three-quarters empty canvas...a form, at once closed and open, full and lacunary, that represented the ideal construction in which to house, with their approximate and disparate localizations, the 'bits' of space that navigators brought back from distant voyages" (*Mapping the Renaissance World: The Geographical Imagination in the Age of Discovery* 7). It was an ideal form for both producing and containing territory. Like all representational systems, mapping depends on the operation of supplementarity, the absence subtending presence. That Antarctica, or the regions of the south, has functioned as the supplement to northern European hegemony is obvious in its very naming as anti-arktos, the anti-north. Early Greek models of the globe have had epistemological and material consequences for the later imagining and study of Antarctica. The Greek T-O globe posited an eternal circle inscribed with the cross of earthly circumstance, leading to the articulation of the earth into northern and southern hemispheres. The Greeks reasoned that a large southern territory was necessary to balance the known territory of the northern hemisphere. The hemispheric paradigm would shape human history in general and Antarctic history directly. In particular, Ptolemy's world map operated like an open canvas, incorporating great swathes of blank space and thus inviting the expansion of knowledge. The renaissance development of a perspectival or holistic globe utilizing the grid of latitude and longitude supplanted the globe broken up into distinct zones. From the modern globe's mathematical projection developed the notion that "anywhere is linked to everywhere" (Agnew), lending a metageographical naturalness to imperialist territorial aggrandizement and capitalist accumulation. As the projected end of the globe imagined as a totality, Antarctica and its mathematically derived center, the South Pole, convened and yet deferred the completion of the globe.

The complex lure of Antarctica as completion of a project of global mapping persists long after the practical attainment of the South Pole, in the cartographic terms of the map, particularly in the Antarctic Convergence, a circumpolar border set at roughly 70 degrees South. Antarctic Convergence as a metaphor insists on the relation and co-constitution of economics, knowledge practices, and materialities of ice. Yet, the Antarctic Convergence also coordinates ecological, cartographic, and biological scales. The Antarctic Convergence is impossible to place precisely. It is a water zone, what is often considered a natural demarcation of flows of temperature, seasonal shifts, and fauna and flora. The longitude and latitude of the grid, on the other hand, derives from the earth's axis, a geophysical referent that does not change, even if the surface of earth does shift as it orbits. The rigidity of latitude and longitude vis-a-vis the movement of earth's crust necessitates the periodic recalculation and replacement of the South Pole marker on the ice surface. Even more, it marks the skewing of the grid. Thus Antarctic Convergence, and the South Pole are lines/zones/points on the terrestrial map that are derived from overlapping systems, one more in relation to astronomic orbits and the other more geo-abstract; one sloshing with celestial and fluid matter, the other a linguistic-cartographic overlay.

Onto this Antarctic Convergence, I layer and plot out convergences of ecological and economic compression and reorganization. Antarctic Convergence becomes a site of "experimental geography" where the rational grid compresses and even collapses toward a more universally imagined geography of the South Pole as the last place on earth. Antarctica and the South Pole developed as place-ideas that were influenced by their geophysical inaccessibility. Such connotations, however, have more to do with coincidence than any necessary properties of place. The deep sea and certain mountainous regions are equally or more remote from human habitations and even less suitable to extended inhabitation. The South Pole as the last place on earth has been invested with the aura of transcendence and promise of the completion of knowledge: Its attainment would close what had been up until that point an open system of global mapping. Inevitably, the physical attainment of the South Pole (and by extension, the continent and Convergence) could never coordinate fully with its multiplying metageographies. In seeking to put an end to the expansion of cultural fantasy around the last place on earth, geographical attainments in fact only led to further attempts, arrivals, gestures of closure, or newly extended limits that could not be contained by their own impetus. These are some of the reasons that I argue Antarctica and the South Pole are distinct from other remote locations.

Antarctic Convergence thus is the potentially open, shifting directionl flow associated with Antarctic geophysics, or what I call the ontology of ice. Destabilizing this anthropomorphism of biopolitical population construction, Convergence situates ice as a material actor, highlighting the disjuncture between the mapped projection and the ongoing disjunction nature of ice and human time. My articulation of Antarctic Convergence resists the very structures that produce Antarctica as the end of the earth and as a permanent and instrumentalizaed "outside" of world systems. "Antarctic Convergence" is also a mode of synaesthesia, or a crossing of human and nonhuman scales and ontologies; it goes beyond making arguments about truth or the real and instead aims to produce change through questioning its own assumptions, methods, and tendencies. Convergence, like the metageography of the earth offered by the "Holes at the Poles" opens and mixes. It counters and ignores origins and ends. It wipes out modernity's precession and its time–space of prolepsis. It is an ecological form of earth, nested, interrelated, and yet open to change. It is not the "whole earth" of 1960s' sentimentalized universalism, but a "shattered" globe of parts, striations; organic fixtures, if you will, of poles that offer direction and once achieved, redirect those very arrivals.

Today, the legal mapping of Antarctica is barely managed under the 1959 ATS. The ATS secures Antarctica as a nonnationalized territory for science. Like other global political organizations and agreements, it is both irrational and effective, meaning it brings together and organizes the need for a global political community and yet does not actually work in literal terms. The legal space of Antarctica produced by the ATS has organized yet irrationalized the territory, making it available to forms of management of territorial claims and occupation, even as it continues to defer legally binding territorial claims into the future, and to allow overlapping claims and even non-claimancy among signatory nations. The next section discusses the confusing effects of the normalization of sector theory on Antarctica's modern map.

Retrofitting the Last Place—The Persistent Failure of Sector Theory

The mapping system that emerged to organize contested territorial claims was the polar sector. A sector is a pie slice of territory radiating north from the South Pole to the ragged and continually calving, thus unmappable, edges of the continent. Using a polar projection centered on the South Pole, the sector map layers the geophysical actuality of the continent's circumpolarity and generally round, contained form (except for the peninsula jutting northward) with a more geometrically regular overlay of the grid of latitude and longitude. The resulting map resembles a pie chart common to business visualization, with the continent as the pie and the sectors like wedges emanating north from the point/center of the pole. This visual coincidence of cartography, geophysics, and board room goes almost entirely unremarked in the naturalization of the map of the Antarctic. But the coincidence of geographic and nongeographic spatial configuration and connotation is actually very significant and accounts for much of the power of this map to effect political negotiations and general knowledge as the sector map illustrates countless articles and books on Antarctica, whether or not they are directly concerned with territorial claims. Although only one form of mapping Antarctica, I argue the sector map is the most powerful and undertheorized representation of the continent that renders Antarctica as excessively subject to rationalized spatiality—or, to being "sliced up"—open to fair and rational competition, stable as territory, and clearly bounded by a naturalized border. This overpowering map has silently restricted geopolitical, aesthetic, and temporal conceptualizations of Antarctica.

The sector map imposes a rigid human-centered temporality upon the complexity of continental ice. All sectors contain base points within their scope. They extend north to the limit of the Antarctic territory at 60 degrees and south to the South Pole. Sectors abstract and generalize territorial claims. Working through whole-earth spatialization to encompass and contain specific and multiple and overlapping national claims, the sector projects backward from the time of the "last place" of geo-historical attainment of the pole, back out north to higher latitudes. In a way, the sector map reversed the time–space constrictions encountered by the explorers as they neared the South Pole terminus of 0 degrees latitude and longitude, expanding out from the pole and recuperating the constrictions of historical discovery in the retrospect of territorial futurity.

Because the sectors organize claim-space retroactively and schematically, they cannot totally account for different types of claims. For example, Norway's claim has no north–south limit and so is not based on the sector. Overlapping sector claims are "layered" on a flat map using dotted lines to indicate disputed or shared "slices." But as in a Venn diagram, which they resemble, these overlapping sectors cannot indicate easily temporal sequence. Equally confusing, then, is the impression that the overlapped subsections of multiple claims are "shared" or common to both national sets, as might be construed from a scan of the map.

The inequalities and unevenness of the widely used sector map do not only extend from its depiction of competing claims, but from its very use of the polar projection in relation to plot-point claims. The "slices" of this cake are not all equal because there is one base claim that does not fall within the radar sweep of a sector. The United States established a base on the South Pole in 1951, at the height of the Cold War. The US Scott–Amundsen base, despite the ATS and its internationalist renaming in 1963, remains a vestige of US post-WWII empire. Given the restructurings of US global power since the 1970s, the sector map is a precious retention of Cold War power that when examined creates a jarring effect with global reality and international treaties, beyond the Antarctic. Most importantly, despite the opposed views of signatory states such as the United States and Australia on the legal force of claims (Australia officially claims 42% of Antarctica while the United States has never filed a claim), the sector map produces the appearance of coherent and equivalent national territorializations. Nothing could be further than the truth.

Centering on the projected "end" point of the South Pole, the offical sector map distorts the actual historical trace of humanity as well as the geophysical actuality of the terrain. It omits and even erases the human history of knowing along with the geophysical features of the vast continent. It compromises accuracy for symbolic efficiency and surface integrity; it replaces complexity with simplicity. Like the mapped globe, the sector map projects a global, total view while obscuring its partiality stemming from its historical context, cartographic templates, and ideological entrainment in the post-WWII power structures of the Treaty System. Alternative mappings of course exist. The next section looks at Antarctica through the promise of GIS as a more accurate and politically neutral mode of mapping.

Google Earth, Antarctica

Open hypertext environments (like the World Wide Web) and "interactive" (relational) environments with transmutational or evolutionary potentials built in need really new virtual concepts. Or new really virtual concepts capable of grasping process unencumbered by reductive spatial or even spatiotemporal framings. They need philosophy. (Brian Massumi, *Parables for the Virtual* 75)

Google Earth Antarctica relies on "collective data collection," to create a more accurate, whole earth. Just as for other sites on Google Earth, individuals volunteer their images for coordination within Google's satellite mapping. This alternative to nationalist mappings, sectorizations, and other overlays that have controlled and distorted perceptions of the continent should provide the most up-to-date map of Antarctica that reflects recent advances in global consciousness and specifically in Antarctic knowledge. Imagine my surprise, then, when I clicked on the ever-changing citizen-participant map. Expecting a more detailed and accurate alternative to the typical white or light blue "empty" expanses of the map of Antarctica, what my eyes found was a mostly white space. Despite the superiority of visualizing and data gathering and the help of countless people on the ground with cameras, Google Earth's polar regions, especially Antarctica, more resemble the nineteenth-century imperial maps that inspired Joseph Conrad to dream of filling in all the "white spaces." Click after click, Google Earth Antarctica produced blank pixels, views that made Anarctica seem preversely less utilized and known than the obsolete Cold War maps in which overt military paranoia populated the poles with enemies, hidden or otherwise, and projections of military bases

(Operation Highjump in 1957 in fact brought the greatest number of people into Antarctica ever) or the continent's present spatialization as the "laboratory for international science" or as anything more promising. Brian Massumi describes exactly the problem of Google Earth Antarctica. Its program is all body and no brains. Or, perhaps, I should say Google Antarctica's problem is in its many *bodies* and overly simplified notion of population, if not populism (Farman 2010).

Google Earth is in a midpoint of utilizing surveillance and disembodied technologies of visualization and editing and yet relying on people/population to produce and upload the data. This is creating a distorted new empire of Antarctica that like the old version of empire remains a serviceable blank, though differently produced. Google Earth Antarctica is a new white blank, the blank of people not caring, not wanting to read the archive; or, the blank that comes when people or population is the measure of representation/visualization: the dangers of biopolitical new media. Surveillance by the people as terrestrially located is not enough to ensure the proper apportionment of attention and resources. If accuracy or scientific concern for global warming were the measure or mode of mapping, probably Antarctica would be the most densely represented landmass (and in the literature of global warming, Antarctica as ice is perhaps overrepresented beyond its capacity to signify as anything other than as data supporting ecological crisis). Instead, Google Earth Antarctica reflects directly its sparse population, a form of reification that strangely mirrors imperial blanking for resource exploitation, or the Bush-era fantasy of wasteland Arctic National Wildlife Region open for exploitation.

Conceptual artist Peter Fend's "Ocean Earth" project is an interesting precursor and alternative to the geographic populism of Google Earth ("Conversation" 2008). For Fend, step one was the economic procurement of the digital data, the satellite images themselves, in the early 1980s. Fend produced a visualization of earth through procuring satellite imaging data through consumer agencies and distributing them though mass media circulation. Fend critiques Google Earth's data collection and presentation as mere visualization, narcissism in pixels that obscure its politics: "Google Earth will not show you what's happening in Iran or what's happening in hot spots" (132). For Fend, the purpose of visualizing data collectivity is to create counter-information and to legally structure rights and responsibilities in relation to the facts of governmental activities across the globe. Creating knowledge about the effects of humanity on the globe unmediated by art galleries and markets, news media outlets, or government control of satellite imaging is another facet of Fend's Ocean Earth project. Although Google Earth seems to offer a way to coordinate terrestrial images through interactive satellite mediation, its populism is an illusion. Google Earth most perniciously manages to close off official and/or traditional forms of protest. Ironically, Fend's Ocean Earth project suggests a collusion of elite art, big science, and corporate strategies of capitalization to counter mindless populism of a neoliberal consumer-driven world drowning in data it revels in producing and amassing, yet has no will to read.

Blank Maps and Black Sites

The map of Antarctica that emerges from Google Earth satellite–citizen nexus may be constructed by technologies of vision and networking unavailable to earlier centuries. Yet the result—a largely blanked out and sparsely represented terrain dotted with concentrations of human-centered built environments—reproduces and even reinforces an idea of Antarctica as a schematic, "ready-made" frontier ever available as a space in which to advance forms of neoliberalism in the face of persistent fantasies about empty territory through a collusion of international science-as-colonization, visual spectacle, and transnational corporatism allied with conservation/environmentalism.

But there is an Antarctica that can never appear on conventional maps or GIS. Already in its short human history, the built environments at McMurdo Sound and at the South Pole have undergone transformations and in fact erasures from sight. Two examples will suffice. The first is what is called "old" pole station. It was built hurriedly as a bunker underneath the ice during the period of the International Geophysical Year (IGY) ostensibly to support US science, but also as a strategic occupation of the most symbolic location on the map—one whose value worked within yet at odds with the non-claimancy and international scientism policies being promulgated by the United States. Certainly for those military personnel who built and lived in the old station, it was a working military installation, despite the official demilitarization that came with the Treaty going into effect in 1961. The ice being what it is, old pole station became impossible to inhabit over time. And anyway, the US science program was growing beyond the capacity of the warren-like dugout. The old pole station was open for visitation until it was deemed too dangerous to allow visits to what has effectually become a monument encased under ice.

Figure 1.1 Connie Samaras, "Buried Fifties Station," courtesy of the artist, copyright 2005. From the series V.A.L.I.S. – vast active living intelligence system, archival inject from film.

Old pole is easy to miss when scanning the ice plateau around the vicinity of the pole. Certainly it does not show up on satellite mapping; nor could users upload their pictures due to restrictions of personnel from its area. Photographer Connie Samaras, a grantee of the United States Artists and Writers Program (USAAWP) in 2004–5, managed to photograph the entrance to the station from the aerial vantage afforded by a weather balloon (Figure 1.1).

The eye strains to locate the tiny dark hole amidst the bright white of summer sun and the pattern of the wind-rugged sastrugis of the polar plateau. But the hole is there, eventually drawing in the patient eye. But to what is the eye drawn, exactly? From scanning a vast field of ice (and within a large photo frame), the eye focuses on a point of nothingness. Like the time–space constrictions of latitude and longitude as they end at the point of the South Pole, the old station is another form of terminus, but instead of being produced by the logic of the grid, it shadows the grid. It is a built environmental overlay on the mathematics of the grid. Yet, instead of yielding a knowable territory, this over-gridding of pattern and vision produces less and less to see and to know. And so, turning north from this foretold end at the South Pole, mystery, the precondition and opposite of science, makes itself known. Samaras's "Old Pole Station" draws in the eye and centers a mystery much like the persistent myth of the "Holes at the Poles": It transforms terminus into potential. However, Samaras founds her new hole at the Pole not on myth, but on recently obscured and constructed (if not exactly suppressed) installations—real material features of the built environment.

The range of visualizations of Antarctica, including mapping and contemporary photography, are entangled in the network of visual problematics that most often have been called "Whiteout." Whiteout is a common and much-written-about phenomenon in polar and other ice environments by which the light overwhelms the visual capacity, creating a monochromatic flooded field in which hue, background and foreground, depth, and all other signals of conventional proprioception is lost (Fox 2005; Noble 2012). But what the trajectory of mapping I have described in this chapter has produced is a distinct form of whiteout that is more a blackout. Whiteout among its many connotations and uses tends to suggest a naturalized limit to human seeing, which is why its effects are most often linked to polar, extreme or aerial environments. Yet, whiteout describes not so much the extremity of geographical environment, but the extreme visual environment. Visual environments are remote, virtual, and disembodied. They are relations among bodies, places, and the mechanics of optics. To this extent, they are not natural and in fact are produced within networks of culture and politics. Blackout, then, names the visual effect at the nexus of optics, the human body, the extreme environment of ice, and geopolitical power.

Another site not available through either conventional or Google Earth mapping is the nuclear power plant built in 1961–2 by the US Navy to serve McMurdo station's energy needs. "Nukey Poo," as the Navy workers named it, or PM-3A, the Martin Co. (early Lockheed-Martin) portable, medium output nuclear plant never lived up to its utopian post-WWII promise to provide cheap and efficient fuel. It was dismantled at great expense by 1972; all its parts as well as the radioactive contaminated dirt around it were shipped back to the United States for containment (Wilkes and Mann 1978). The plant once stood part way up "Ob Hill," the major geophysical and cultural feature of McMurdo Sound, from whose summit the British explorers of the Heroic Age surveyed the area and where today a cross erected commemorating those explorations still stands. In fact, that commemorative wooden cross battered by the elements has been refurbished numerous times and remains the destination for visitors and workers alike, who enjoy a quick climb to pose for "hero shots" at the summit (Spufford 2005). But the Ob Hill monument that no one can visit and that has not even a plaque to mark its having been there, is Nukey Poo. As a product of the built environment that is no longer available to sight, it is what I am calling here a site of "blackout" after Trevor Paglan's "Black Sites"—photos, geographies, and people that "don't exist." In related work of satellite-informational excavations of desert sites such as The Salt Pit (2006), Paglan uncovers "geographies of places outside sight." Paglan's photos become "proof" of what is not there, no longer there, or what never was there. They are the obverse of Google Earth's mapping.

Because black sites are constituted in networks of secrecy, in desert terrains remote and otherwise difficult for human embodiment, and as optically impossible objects, their "discovery" requires the use of the very means through which they "disappeared." Moving through the sector and grid, Paglan utilizes the techniques of biopolitical control, cross-referencing databases to locate the missing places and people, the "ghost" prisoners and hidden sites. He ironically uses a gridding technique to yield the holes in the mesh, the absences constituting the "outside" of the grid, the points between the points. These non-sites or non-places destabilize distinctions of inside/outside, seen/unseen, knowable/unknowable that found rational conceptions of space. Geo-artists like Fend and Paglan use controlled satellite images and redistribute or recommit to oppositional informatics to produce counter-knowledge, the very absent spaces blacked out by human experience and memorialization and posthuman technologies such as satellite images as controlled by government agencies. There is no grid of ecology/knowledge to tear a hole in. Relying ultimately on a retroactive human presence and an inadequate notion of visual accuracy across incommensurate regions of the earth, Google Earth tries to fill in holes only to end up revealing that much more is missing than geographical coordinates. We are still missing a philosophy.

Chapter 2

Refusing History after Ursula K. Le Guin's "Sur"

The "Heroic Age" commenced with the revival of interest in Antarctica around 1900 and ended before the era of mechanization and aerial reconnaissance beginning in the late 1920s. Its colonial modes of knowledge were limited by the ability of the human body to withstand the environment. Human inhabitation was its goal, experience its method, and the footstep—the mark of embodied movement and progress upon the terrain—became its prime symbol. The material facts of the race to the pole are as easy to trace as tracks in snow. Amundsen dog-sledded from the Bay of Whales beginning on October 19, attaining the pole on December 11, 1911, and returning safely to base. Scott man-hauled his sleds, attaining the pole on January 17, 1912, five weeks after Amundsen. The British team was caught in a blizzard 20 miles from a food depot. The entire polar team froze to death in or near their tents. Members of the base team located and removed the corpses the following summer. Scott's diary was found with his body; his last act had been to write in it. The British have managed to put a heroic spin on an expedition for which heroism and failure seem inseparably linked.[1]

Ursula Le Guin's 1981 short story "Sur" exploits such ironies in the official history of the "conquest" of the South Pole by inserting a women's expedition temporally between Scott's initial 1902–4 expedition and his 1911–12 failed polar attempt.[2] "Sur" stands at the center of feminist and postcolonial critique of polar exploration; it has been almost embarrassingly generative for the project of this book. Not only did Le Guin's near-hoax change the view of what came before it, this chapter demonstrates the legacy of its discursive and political predictions for earth—and ice.

While narratives from official Antarctic exploration focused on the men, techniques, and hardships of polar exploration, Le Guin's fantasy raises issues about feminist historical revisionism. Her chief premise—what if South American women had discovered the South Pole?—projects, as Scott put it, "very different circumstances" onto Antarctic exploration (Scott 1913). While Scott originally used this phrase to reference his disappointment at his belated arrival at the pole, "very different circumstances" opens up for Le Guin a broader arena of competing, multiple histories for any single event. Most crucially, "Sur" presents the possibility of non-European women's prior claim to historical knowledge, considering them as actors in history even when they do not appear recognizable as such in written annals. Much debate now rages over the past of feminism, over the probability of alliances and the status and composition of originary feminism, with some factions (or individuals) claiming the movement and others claiming to have been written out of accounts of its founding. Le Guin's destabilizing fiction has profound implications on any feminist futures being presently produced. In particular, it complicates accounts of feminist history that omit subalterns, especially if they are not accounted for originally. Le Guin's suspicion toward autochthonous ontology—or narratives which place their authors at their own beginnings—might prove productive for feminist, subalternist, and postcolonial theorizing both against the presumptions of standard history and within competing revisionisms. My analysis of "Sur" focuses on the tension between the possibilities of reform and the techniques of repetition, posing such questions as: Do the women demonstrate a path to the pole that is distinct in intent, execution, and effect from that of the men of history? Can Le Guin project a feminist fantasy of prior arrival and yet refuse the complicities of a masculinist teleology of conquest? And—to extend the historical metaphor at the heart of "Sur"—can feminism discuss its own pre- or nonexistence and still, to echo "Sur"'s last line, "leave no footprints"?

My investigation of the politics and effects of Le Guin's retemporalizing of the exploration history of Antarctica is organized into three parts: First, I demonstrate Le Guin's engagement with the archive of European "Heroic Age" exploration narrative, focusing on the footprint as a symbol of Western civilization's quest for total global knowledge. The second section discusses feminist and postcolonial historiography in relation to such foundational history. In particular, I move from acknowledgment of the risk Le Guin as a first-world author runs of merely reinscribing a global feminist discourse antipathetic to subalternist politics, to examine how feminist and subalternist historiographic impulses confront their own limits in the gesture to become both foundational and global, or, to be both origin and end of a teleological process. From the vantage point of the shifting contemporary, Le Guin's feminist parable of 30 years ago evinces not only the temporal disunity within total global knowledge, but also feminism's own struggle with origins and originary acts. The final section takes up Le Guin's ecofeminist themes as they extend through postcoloniality to confront the "hard limits" of the end of oil capitalism and the crisis of climate.

"Sur" parodies the genre of the narrative of exploration, as its subtitle, "The Summary Report of the Yelcho Expedition to the Antarctic, 1909–10," suggests. This report carefully constructs a realistic account of a discovery of the South Pole by a group of nine upper-class, adventure-seeking South American women, who consciously follow the example of the established polar explorers Amundsen, Scott, and Shackleton. With the aid of a secret male benefactor, on pretense of a pleasure cruise, the group sails south from Chile to set up operations at Ross Island on the Antarctic peninsula, previously the base of the Scott expeditions. A smaller contingent of four decides to search for the pole. Hauling their own sleds, the women reach the pole on December 22, 1909, two years before Amundsen and, of course, Scott. They are picked up the following summer by the crew of the Yelcho. Upon returning home, the women pack away the diaries of their exploration and discovery and continue with their conventionally domestic lives, never to announce to the world their attainment of the pole. While the narrative contains suggestions of having been edited at least once through the years by the narrator herself, it remains conjectural by what means the narrative made its way from an Argentine attic circa 1910 to the pages of one of the world's best-known forums for new fiction in 1982.

This secret history revealed too late to affect the course of history enables Le Guin to divert the path of the Anglo-American narrative of exploration from its traditional celebration of the masculine achievement of a geographic goal toward a critique of the problematic legacy of the Heroic Age.[3] Even minor details suggest the depth of Le Guin's critique of European northern hemispheric domination. For example, the ship that brings the women to and from the pole is named Yelcho, whose Argentine captain Luis Pardo really did rescue Shackleton's stranded men in 1914. Le Guin thematizes the northern versus southern geopolitical power struggle over Antarctica. "Second-world" southern hemispheric nations like Argentina and Chile, although closer to the site of Antarctica, were largely written out of the imperial narrative of exploration and discovery. But these nations were not without their own history of colonial interest in Antarctica. It is into this extended history of geopolitical struggle over territory in which first arrivals, territorial possession, and super-session have played a major role that Le Guin inserts her own claim to Antarctica's "dubious capes and suppositious islands" (257) and symbolic landscapes. And like the claims made by competing nations, Le Guin founds hers in the well-established practices of expedition and publishing of the Heroic Age.

Central to these practices are the journal as literary and scien-tific record, and no polar explorer's pen was more important to Le Guin than that of Scott. "Sur" is overwritten by Le Guin's fascina-tion with the narrative trajectory and exploration details provided by Scott's two published accounts of his time on the ice: *The Voyage of the Discovery* chronicling his 1902–4 expedition and his posthumous *Last Expedition* (1913), his journal account of the doomed 1911–12 expedition. These books document the paradoxical aloneness and collegiality of Antarctic exploration. Throughout the Heroic Age, the vast surface of Antarctica was marked with the sled and ski tracks and boot prints of the men who explored south to the pole, and then retraced their trails north to report their discoveries. British, French, Russian, and American expeditions vied for geographic and scientific knowledge of the last continent. Reading Scott's narratives, one finds that this continent, seven times the size of Texas, was seemingly so crowded that the explorers inevitably sighted (cited), intersected with, and traced each others' paths in the ice. Paradoxically, in this most extreme wilderness on earth, explorers relied upon and expected to encounter the marks of other human beings, and the sighting and supersession of the marks of previous explorers were duly reported in expedition journals. Scott, for instance, refers to the "benefit which we had derived from studying the records of former Polar voyages," projecting himself into these records by asserting in his own narra-tive that "the first duty of the writer is to his successors" (*Last Voyage* ix). As explorer, researcher, and writer, Scott envisions exploration as patriarchal paternity, a legacy of connecting marks between successive individual explorers, each of whom contributes through repetition and individual advancement toward final knowledge of the globe.

In the years since it first appeared hoax-like in *The New Yorker* without the authorial byline, "Sur"'s careful reliance on the narra-tives of the Heroic Age explorers, its ecological motifs, and its South American relocation of European imperialism reassert Le Guin's pre-science. Postcolonial and neoliberal political struggle continues to be a feature of Treaty negotiations; women have changed the everyday life in stations, first the ozone hole and now global warming has created an ecologically fragile Antarctica and pressured the ethics of continuing development on ice. But most significantly, as tourism and especially extreme adventuring rise in appeal, the heroic legacy that is both followed and examined by Le Guin has become not merely a backdrop but an unevenly reanimated set of acts, built environments, bodily performances, and narratives whose combined forces have interlarded and become one with the ice itself. Katha Pollitt in her *To an Antarctic Traveler* (1982) called Scott "that valentine thrown out," teasing the postcolonial revisionism that had demoted Scott from Edwardian keeper of the flame of British empire to a figure of revilement and ridicule. Scott's legacy keeps drifting with the ice sheet—literally, as his buried corpse (along with those of his team) remains encased in the ice and will inevitably be ground out to sea, Pollitt points out, with a non-heroic "splash." Imagining Scott as a rejected suitor also evokes a feminized audience for the romance of the south—a demotion in cultural status that ironically dovetails with feminist postcolonial critiques of imperial legacy. Scott keeps moving, with the ice. And like the ice that he has become, Scott still matters. Le Guin knew that in 1981 and the world is still following in her footsteps. Accordingly, the next section contextualizes recent debates around Antarctica's heroic legacy in order to shift, with Scott—and the ice.

Absent Natives

One indication of the need to reassess the so-called race to the pole is the endless inter-referencing books about Scott and Amundsen. Though books about Scott continue to outnumber those about Amundsen, the field can never return to the time before the "rediscovery" of Amundsen marked by Huntford's dual biography, *Scott and Amundsen* (1975), in which he extended postcolonial critique to polar history. Huntford recast Scott, the hero of WWI and beyond, as an inept bumbler and elevated the dismissed Amundsen as a type of the technological traveler and furthermore a victim of British cultural imperialism. South polar enthusiasts still pick sides, Scott or Amundsen. The restaging of the race between Amundsen and Scott in numerous museums and legacy venues as the hundredth anniversary approaches inevitably leaves out alternative histories of race and gender, even by including them within a European frame. Playing on the exceptional nature of Antarctica's originary lack of native population, this section points to the "absent native"—a critical companion to Le Guin's inserted South American woman explorer—to elaborate the continuing effects of European and humanist historical framing of Antarctica's exploration history.

Race and gender are present through their absence in polar exploration narratives in a September 2007 review in the *New York Review of Books* by Al Alvarez, "S and M at the poles." Alvarez discusses David Crane's *Scott of the Antarctic: A Life of Courage and Tragedy*, Sarah Moss's *The Frozen Ship: The Histories and Tales of Polar Exploration*, and Simon Nasht's *The Last Explorer: Hubert Wilkins, Hero of the Great Age of Polar Exploration*. Alvarez reviles the decline of Scott's reputation, the way the "beau ideal of English Chivalry...became a by word for bungling incompetence." Alvarez quotes Crane's defense of Scott at length:

> How has a life that was once seen as a long struggle of duty been transformed into the embodiment of self-interested calculation? How has the name of the meticulous and "cautious explorer" his men followed become synonymous with reckless waste? How has the son and husband his mother and wife described become the type of emotional inadequacy? By what process does tenderness for animal life become a pathological disorder?...If Scott was once celebrated as the incarnation of everything a Englishman should be, he is now damned as the sad embodiment of everything he actually was. (241)

Old imperialists, it seems, are the safest targets for politically motivated revisionism.[4] The question of Scott's radically changed legacy begs the answer that it is not Scott that has changed, but the times. While Scott once got a pass in a way for having used Antarctica, mere ice, as his backdrop, the very starkness and inutility of his arena continues to cause Antarctica as ice to fall away and brings into sharper focus the ideological significance of his endeavor. What has been labeled multiculturalism and its discontents is what Alvarez detects in the scramble over Scott's legacy. Thus, he accuses a less hagiographic account of Scott by Sarah Moss of a "distorting...political correctness" and "shrewd distaste for imperial baggage she believes Scott took with him to the pole." But Alvarez has more on his mind than either an emotional defense of the man or an nostalgia for empire. He moves from garden variety sentimental conservatism to place the controversy over Scott's legacy in the context of US militarized geopolitics, noting "Scott [has been] attacked from both left and right as though he were the General Haig of Antarctica." In this comparison, Alvarez cuts through all the minutia of expedition preparation and decision-making, to suggest that the growing industry around heroic legacy is more properly directed to the broken attachments among the military, masculinity, and the government. Like Haig, Scott had been a representative of a national military system that no longer commanded uncritical allegiance based on its imperial might. Crane's and Alvarez's defense of Scott are symptoms of contemporary neoimperial struggles over the seemingly benign conclusion to Europe's imperial era on ice.

Barry Pegg's article, "Nature and Nation" epitomizes the reassessment of British polar endeavor using race and empire as key terms. Pegg creates a hierarchy of the "interplay of nature and nation" in which Scott shows the "failure of blinkered nationalism"; Peary is higher up, as he followed Amundsen and Cook's appropriation of the Eskimo in the Arctic; and Amundsen is the ideal, combining "nationalism with receptivity to information from others." Those others are his fellow north European explorers, especially Nansen, as well as those "other others," the Eskimo. Pegg ultimately sites lack of "material culture" as a source of failure for Scott and Franklin (215). "Material culture" is a code word for native knowledge. Pegg's reassessment of the heroic explorers is in keeping with the now familiar critique of the complicity of imperial travelers and explorers with colonialism and the destruction of native habitats and customs. However, the results of Pegg's method are far from liberatory or even accurate. In attempting a resurrection of native knowledge within a discussion of how "British, Norwegian, and American culture withstood and interacted with an indifferent nature and uncomprehending indigenous culture" (214), Pegg traps the native within a material culture discourse, while the battling Europeans retain the foreground of historical significance. Europeans again become properly and fully agents of history while natives in such phrases as "the natural world and its inhabitants" are aligned with nature to their detriment. Pegg seems to be protecting the knowledge claims and status of natives—but only to the extent they were instrumentalized or refused by the historically present Scott and Amundsen.

The race to the South Pole is now emerging as not merely the symbolic spectacle between white men revived for a neoimperial era, but as a battle over the absent non-Western polar peoples. One of the obvious calls in light of these critical problems is a need for an index of shifting racial discourses since 1911. Reassessments of Scott and replacements of Scott with Amundsen reveal new racial ideologies that nevertheless continue to recuperate an imperial project, especially in Antarctica, where the lack of indigenous peoples can cover over or distract from racial discourses that operate without natives: revealing the reliance of white masculinity on a serviceable notion of racial difference, even and especially at the very end of the earth. If Pegg followed through on his reassessment of Scott and Amundsen, he might point out that it was the natives, absent at the South Pole, who deserved the prize, or who should today be acknowledged as the proper winners of the race. But neither Pegg nor Alvarez is able to question the goals or the concepts of the "race" or to detach from the task of reviling or defending the white men endlessly within the frame of history.

Alvarez's review concludes with a positive write-up of Nasht's new book on the Australian explorer Hubert Wilkins, a "wild colonial boy who couldn't resist a challenge and went exploring." Here, the conservative reviewer Alvarez and the liberal critics of Scott collude in producing a new white hero from the demise of the Scott legend: first Shackleton and now Wilkins, a "colonial" yes, but also a member of the white settler class of a nation that incidentally claims 42% of the Antarctic continent, some of it based on Wilkins's exploits. Ernest Shackleton's resurgence in the 1990s may have outpaced both Scott's and Amundsen's popularity. His complete divorce from history and utility of any kind—he is lauded only for not dying and interesting now for not being historical and for missing WWI, something that did not go over until WWI no longer rung even a nostalgic bell—makes his story the more complete fulfillment of the nineteenth-century ideal of pure, male suffering. Popular narratives of exploration, it seems, must be continually purged of their links to empire, of their repressed and even absent native cultures. They must become increasingly virtual, that is, based on an original or real that never existed, that is gestured toward, and yet erased in the gesture. Ever proliferating in-the-footsteps-of style adventures partake of this paradoxical refining logic: The more they imitate their heroes, the more they erase whatever history those original explorers represented.

Furthermore, narrating the race as a clash of cultures rather than as a consolidation of European power entrains contradictory fields: It distinguishes among Europeans; reestablishes the very notions of imperial power we attempt to reassess, especially by justifying the winners and losers rather than questioning the race itself; and over-produces a drama whose end is known but whose meaning should not be secured. In particular, it has not allowed the South Pole story to have a relation to present global realignments, especially the neoliberal remilitarization of the poles and the continuing exclusion and suppression of less powerful nations (not to mention nonnational entities) in Antarctic geopolitics. I want to reintroduce a polar history adequate to the needs of the present, and to those peoples and powers that have not already left their marks.

And here, we can return to Scott's belated arrival at an already-discovered pole. There was little to do in the disappointment but to use the equipment they so laboriously dragged with them: they took photograhs that would be revealed to the public many months after their belated arrival and even after the news of their eventual deaths on the return from the pole. The time lag between the taking of the photos as evidence and the making of the proofs is a more material echo of the shadowing and repetition of exploration history. Those photographs first taken at the pole and recovered in 1913 demonstrate the sequence of marking ice. First, the photographs as forms of proof of human being at the place of the pole, and then their promiscuous circulation. But in that space between Bowers's sighting and his knowing; the taking of the photograph and its proof; and between Amundsen's and Scott's arrival is the loss of indexicality, or of the direct relation between actor and mark. The indexical mark of the British at the pole is not broken. The photographs, letters, and acts do not point directly to human agency. Rather, the break in indexicality of these images and acts enters through the desire of the British to disbelieve their eyes. This refusal to accept the conditions of and on the ice, whether of vision, of the acts of men, or of the ice itself, necessitates the endless replaying of a history that represses originary colonialism and produces a nostalgia for the seeming certainties of territorial, indexical empire that ice has never properly supported, and indeed has undone.

Imperial geography was an elaborate act of leaving of marks—footsteps, equipment, and letters—for an audience without whom the very act of achievement would have very little meaning. The problem of leaving marks, or of being historical in the highly overdetermined place of the South Pole invokes basic questions on the nature of history, both how it is constructed and to what ends. Scott's "different circumstances" hint at the conflicts of perception, timing, and of national interest within the seeming availability of the blank whiteness of Antarctica. These overlapping, canceling, erased, missing, and contested footprints are the strange indexes of humans on ice. What are these footprints still being left and followed on the ice? A footprint is an indexical mark of a creature's passing. Or, the impression of light on photographically prepared paper. The imprinted light of photography in particular forms a crucial archive that recently has been enhanced and challenged by digital technologies. These marks may be absent or missing, or take the form of absence or supplement. Certainly, these marks are not strictly visual. Memory and affective flows of belonging and most crucially anthropogenic practices contributing to warming take place within scales and ecologies beyond the human sensorium or representational field, or to ice itself. But the way fields and questions are framed also leave marks.

The Problem of Gendering Antarctica

One of the most persistent myths of Antarctica is that in its Heroic Age, it was an exemplary masculinist space. While even that observation is not a simple one, and it has not always been treated simply, the naming by feminism of Antarctic history as masculinist has been less accurate than perhaps obvious. In decrying perceptual and actual patterns of empire and imagination as masculinist (not to mention white), feminists have in many ways foreclosed alternative histories by reifying the past as immovably and irretrievably masculine and European. What I mean is, that a certain type of critique, of pointing out the terrain overrun by men and male culture (whatever that is) misses the women in the landscape. And by this, I do not mean the many personifications of ice as a cold female lover efficiently parodied by Pollitt's comparison of Antarctica as "pale Narcissus" and "Pavlova...turning" in self-regard, and finally as "cold Versailles," an uninviting house of mirrors.

Rather than these gendered myths of Antarctic ice's remoteness and thus its availability to a form of romanticization and to being inevitably conquered, I mean something more like the story I got from Gerda Lerner, a foremost feminist historian of the positivist/revisionist school. When I mention my research topic to most people, they respond with curiosity and surprise, and pepper me with questions like "How cold is it, really?" They have very little facts and even less imagination. Yet, that is not what I got from Lerner. Instead, she related that one of her most vivid memories of her Hungarian childhood was of reading Ernest Shackleton's *South*, the account of his 1914 attempt to traverse the continent. His ship became iced in, and the journey into history took a turn from geographical feat to elemental survival. It took 60 years for Shackleton's failure to become re-heroicized as a great tale of survival. But to Lerner, the details of the preparation and their thrilling possibilities remained clear in her memory. Shackleton's famous (and as it turns out, mythical) advertisement "Men wanted: wages small, safe return doubtful" hailed many girls and women for whom identification with the exploration journey to Antarctica was produced by its very impossibility. The homosocial directive of Edwardian exploration culture is rendered a humorous set piece in Friend's 1948 *Scott of the Antarctic*. In one scene, a young girl reports to Scott to be considered for the expedition. Rather than dismiss her outright, the film's kind-hearted Scott humors the patriotism of the eager girl while deflecting her to a more gender appropriate support role—indeed, the role of the vast majority of the public. Categorically shut out of the advertisement for men, girls and women nevertheless were to make an imaginative place on ice as a platform for grand imperial endeavor that was not directly military in nature; in fact, the Antarctic expeditions of the Heroic Age tended to be funded largely through direct subscription, not through governmental channels. For these reasons, Antarctica was the first territory open to female imaginaries, not so much because of what it was (remote, wasted), and not even because of what it was not (national territory, inhabited), but for what it could be: a world even men could not make their home. Because of its being left out of the normal circuits of value produced by the militarized subjugation of native peoples, Antarctica was a perfect location for the contradictory expression of the desire to colonize from the point of view of those already left behind, at the center of empire.

Complicity in empire, or the imaginary desire for a heroic position and power, even if only through heroic narrative, cannot be the end of the road for thinking through feminist encounters with polar territories marked as terminally masculine. Given the impossibility of engaging Antarctica outside imperialism, what kind of Antarctic place might a feminist postcolonial political horizon impose? To continue to consider this question, I will turn to Antarctica's most celebrated contemporary cultural historian, Stephen Pyne, author of the 1987 book *The Ice*. More recently, and perhaps reluctantly, returning to writing about Antarctica after building a career as an environmental historian of fire, Pyne has argued that due to Antarctica being "inextricably different" ("Extraterrestrial" 147) in its lack of sustenance for human life, what he terms the "Third Age of Exploration" represented by Antarctica and outer space will not fulfill dreams of new land presupposed by previous eras of exploration; instead, it will produce mere mining town-like outposts, but not a true civilization. Antarctica's Third Age of Exploration will be more like exploration's past, Pyne argues, characterized by high-investment forms of "luxury" data production, which he likens to the spice trade of the Second Age, and by robotic and remote technologies to replace human inhabitation. Pyne's dismissal of any utopian delusions around permanent or large-scale human colonization of Antarctica aside, of even greater interest are the spectacular moves and erasures necessary to Pyne's undermining of progressivist history.

Pyne's critique of international science policy in Antarctica is marked by a nostalgia for first-order imperialism, as compared to the Antarctic case: "Those [meaning white European men] in those outposts could intermarry among indigenous peoples, and the resulting mestizo societies...did the heavy work of exploring and settling the interiors" (148). Such pining for the days of the creation of an entire civilization of subalterns leads directly to Pyne's seemingly total refusal of empire in listing features of the contemporary Antarctic such as "tourists, Chileans, quasi-military social order" (148) but omitting the most obvious human permanent culture, the US bases. This omission results in the discordant vision of the Third Age as "[absent] of the moral conflict that so stained European expansionism...uncontaminated by imperialism" (148). American exceptionalism, or the view that US historical formation was in fact extricable and from that of Europe, and a view of Antarctica as radically different from the rest of the earth convenes yet another discordant lament about the "price" for the "blessing" of Antarctic endeavor's moral purity being an "absence of moral drama"; Antarctica is disappointing to a grand narrativist as it cannot extend the stories that once produced it as the terminus and goal of exploration. Pyne refuses to encounter Antarctica as rupture, though he gestures to its limits and disappointments for a specifically human belonging.[5] Pyne ends this passage in high symbolic nationalism: McMurdo [the US base situated on McMurdo Sound in Antarctica] is "not Plymouth colony but a Virginia City," suggesting that Antarctica is a failed frontier. Yet, we must wonder why a Third Age when the first two led to this intractable end? And does he really think the United States has not colonized Antarctica with its ever-expanding bases and operations? Pyne sends a mixed message to those looking for new lands to colonize: Antarctica will not become that place, even as he displays an odd form of desire for the previous ages of exploration and colonization. It seems that Pyne must condemn Antarctica's potential as a site of colonization even as he refuses to imagine this history differently.

How does a feminist imaginary—ironically structured similarly to Pyne's argument in its cleaving to a history of expedition and colonial fantasy only to refuse it—find a path that can be maintained to understand humanity's claim upon and responsibility to the polar regions generally and Antarctica in specific? Postmodern feminisms (and other isms) can study Antarctica as a place that is shifting as it is being studied; absolutely constituted under imperialism; and perhaps best served by a "clean-up"—or rehabilitation of the imaginary as in the discussion over what to do with Scott's hut in "Sur." A feminist utopia creates an alternative path: a post-neoimperialist move to wipe clean the dirtied *tabula rasa*. However, as the next section demonstrates, feminism as a postcolonial condition inevitably retraces the history it critiques.

Millennial Antarctic Histories

Capitalizing on millennialism and in anticipation of the hundredth anniversary of European men reaching the South Pole, in October 2000, two women began a sledding expedition to cross the Antarctic continent without mechanical aid. Their goal was far from a first. Ernest Shackleton first attempted the feat in 1914. After his ship was crushed by the ice, Shackleton became famous not for crossing the continent but for simply surviving. The men of later eras successfully endured crossings both mechanized and on foot and ski. Other Antarctic achievements—circumnavigation, overwintering, and cross-continental overflight—have all been accomplished.[6] So why, at the outset of a new century, were these women pulling sleds across the vast and inhospitable ice, when continental crossing had been repeatedly achieved? I want to approach this question by placing this most recent example of repetition compulsion at world's end in the context of a larger concern with the problem of "making history" for feminism and, in a related sense, for postcolonial studies provoked by "Sur."

The title "Sur" suggests layers of meaning for feminist and postcolonial discourses. "Sur" is Spanish for south, linking with singular economy Le Guin's indebtedness to normative history and her intention to rewrite that history, or better, to reroute its focus on first-world centers of culture and on European agents to consider history from the perspective of its southern hemispheric other. "Sur," or south, corresponds to Antarctica, the seventh continent, a place devoid of indigenous human population and almost entirely ice-covered. While occupying an enormous landmass, Antarctica has nonetheless remained marginal at best to the struggles of humanity. Le Guin chooses Antarctica as setting for her critique of a European-dominated history not in the least because of the territory's lack of direct significance in that very history. Despite the attempts of explorers and writers to interpolate the frozen continent into the annals of imperial conquest, Antarctic exploration has held a largely symbolic value and those who have pursued it have done so with minimal government support. It is this vexed temporality—where origins are never actually firsts and resistance must become originary—that the women who most recently crossed the ice in 2000–1 inhabit. That their inhabitation of Antarctica is less about space or the materiality of the continent than it is about inhabiting time is one of the central issues this chapter considers.

Both feminism and postcoloniality have developed as critiques of a historical narrative that finds them ontologically belated. Feminism is often understood as the movement produced to counter a masculinized historical normativity, much as the postcolonial arrives late—after the breakup of empire—to both political consciousness and knowledge production.[7] The impact of these critical discourses arises from their complex relation with the histories which cast them as coming after. Feminism often pursues foremothers to balance the temporal preeminence of founding fathers, while postcolonial studies argue for the power of precessionary indigenous civilizations and subaltern subjects. The effect of such work has been somewhat paradoxical: both feminist and postcolonial studies resist dominant discourses by inserting alternative models for knowledge production, yet they can also reify dominant history in their reliance upon origin narratives as the source of their critique. This tension between complicity and resistance is a feature of the retemporalizing historical critiques of each.

The problem of coming after, or of belatedness, is thus built into the critical projects of feminism and postcoloniality. But any first, even the first discovery of the South Pole, actually comes at the end of a series of events; both the territorial space occupied by Antarctica and the timing of its placement in history illustrate the ways that beginnings are also ends. Antarctica's discovery symbolizes and makes concrete the end of an extended history of Western civilization's exploration of the globe. The achievement of its most geographically remote location, the South Pole, in the years between 1900 and 1912 constitutes the "Heroic Age" of Antarctic exploration and coincides with the rise in Euro-America of a modernist program aimed at total global knowledge.[8] However, the sense of finishing the globe has been recast at the turn of the twenty-first century as a story of environmental degradation on a planet made smaller by the accumulation of knowledge in excess of its object. Antarctica in the twenty-first century is thus a symbol of twinned hope and despair. On the one hand, it symbolizes desire for an uncomplicated human capacity for advancement and ingenuity. On the other hand, its metageography as the last place on earth represents the limit of possibility for colonialism: It is a frozen wasteland that according to Western notions of embodied presence in real time cannot be profitably inhabited.

Apparitional Footsteps

Le Guin's powerful revision of exploration history in "Sur" arises in part from her ability to interrupt patriarchal succession by crafting her narrator, now a grandmother, as simultaneously the successor and predecessor of Scott. Troping Scott's detailing of the techniques and supplies necessary for polar exploration and of Antarctic geography, Le Guin presents the women hauling their sleds, building supply cairns, taking sightings, and becoming snow-blind and exhausted in much the same way as did Scott and his men. As a member of the female exploratory team exclaims, "[I]f Scott can do it, why can't we?" (258). For Le Guin, however, the ability to follow in Scott's tracks is only valuable insofar as it engenders a critique of the idea of heroic arrival. The last words of the anonymous narrator's exploration journal read, "we left no footprints, even" (270). In her character's refusal to leave a readable trail, Le Guin undoes the teleology of masculinist exploration in which explorers reproduce themselves by beckoning the arrival of their successors. The question is, does Le Guin foreclose a future for feminist and subaltern resistance by inventing what some historians would define as an "ephemeral" archive?[9] If the women's expedition does not leave footprints, then what can it leave as a history? How are we to understand the possibilities of a history so unreadable by a world outside the restricted "family" circle of the subaltern actors?

The metaphor of the footprint, the mark across the ice, stands also for the mark across the page, particularly those references so aptly called footnotes that represent the construction of a deed in the context of history. In declining the exploration protocol of publishing their journals, the women leave neither narrative, footnotes, nor footprints. Le Guin, nevertheless, employs the central metaphor of the footprint to suggest the complexity and significance of the decision of her narrator to leave her mark neither on the continent nor in history.[10] The South American amateurs discover a set of "footprints standing some inches above the ice" on the high polar plateau. The narrator explains, "[i]n some conditions of weather the snow compressed under one's weight remains when the surrounding soft snow melts or is scoured away by the wind; and so like reversed footprints [these prints] had been left standing" (267–8). Here Le Guin picks up Scott's trail, for her narrator's description of these unusual footprints stems directly from Scott's discussion of this distinctly polar phenomenon and his repeated deployment of the technique of tracing the trails of the outbound party by the returning party. The uncanny ability of the Antarctic environment to preserve even the most insignificant human trace was often noted in expedition journals.[11] These apparitional, disconnected footprints symbolize misgivings about the purpose, methods, and goals of exploration as history. Not only do the women choose in a variety of ways not to follow in Scott's footsteps, but also the symbolic footprints of history stand reversed and disarticulated, leading nowhere.

Following Scott's footprints, moreover, would have been impossible, since the land mapped by Scott in 1902 had already changed. The women come in sight of the "Barrier at the place where Captain Scott's party, finding a bight in the vast wall of ice, had gone ashore" (265). They discover that the "sheer cliffs and azure and violet water-worn eaves" of the Barrier were "as described, but the location had changed" (265). The unstable ice shelf has deranged the landscape of Scott's narrative/map. If the women had planned to use Scott's journal of the area to navigate their own way to the pole, they would find the expected landmarks askew. Scott described his Antarctic; the women must find their own way on the mutable ice. And yet, the women's progress to the pole is measured against that of the "real" history of which the reader is well aware. Not only do Scott's journal entries provide reference points for the fantasy expedition, but on the all-preserving continent itself the women discover archaeological remains. This phenomenon of preservation occasioned a journalistic outpouring when Scott reconnoitered his former hut built during the 1902–4 expedition. In 1911, Scott found his formerly snug habitation in a state of preserved ruin, filled with snow and unusable as shelter. He is "depressed" by the "desolate condition" of the hut, and the lack of civilization evinced in the neglecting of such a "simple duty ... by [his] immediate predecessors" (*Last Voyage* 95). For Scott, the preparing of a readable text for his successor is as important as the proper upkeep of the hut. To Scott, the hut represents England, home and civilization, an oasis on an unfamiliar landscape, and its maintenance emblematizes colonization. Scott's sense of his monument having been defiled, and of his subsequent spiritual "oppression" is recast by Le Guin in her own hut-reconnaissance scene. The text of the hut provides an opportunity to critique the methods and end effects of imperial exploration: "[T]he large structure built by Captain Scott's party stood, looking just as in the photographs and drawings that illustrate his book" (261–262).[12] The women approach the abandoned wood structure, artificial against the background of ice and rock, and surrounded by inquisitive penguins. They encounter a sight, to analogize from colonial history, of a former settlement, deserted and overrun with natives/penguins. The hut functions (for Scott as well as the women) as a memento mori of imperialism, a reminder of the passing of human endeavor at the very scene of its self-defined triumph.

Although the history and dignity of the lone outpost impress the women familiar with its aspect from Scott's journals, their changed circumstances displace the written image. Placed in the women's cultural context, the hut represents something more deeply inadequate than a defiled haven. The narrator continues her description: "The area around [the hut]...was disgusting, a kind of graveyard of sealskin, seal bones, penguin bones, and rubbish, presided over by the mad, screaming skua gulls" (262). The "slaughterhouse," as another expedition member refers to the hut, represents a hell on Antarctica created in the wake of Scott's environmental disruption. Le Guin's explorers invoke an image of apocalyptic renewal when during a discussion of how to use the abandoned hut one woman proposes that they "set fire to it" (262). The women find the interior of the hut to be less offensive but very "dreary": "It was dirty and had about it a mean disorder. A pound tin of tea was standing open...a lot of dog turds were underfoot—frozen of course, but not a great deal improved by that. But housekeeping, the art of the infinite, is no game for amateurs" (262). Beneath the humor of the genteel adventuress' disapproval, Le Guin embeds a more serious message for her 1982 audience witnessing, among other growing disasters, the destruction of the ozone layer above the South Pole.[13] "Housekeeping," in the context of the women's world (and that includes the far reaches of Antarctica), is a prime goal in an environmentally conscious age. Incidental to the women's refusal to leave marks is their refraining from leaving garbage. Le Guin's description of the women's camp is the story's most radical departure from the realistic imitation of historical exploration narratives evident in the story as a whole and shifts into a frankly utopian modality, invoking a recognizable US-style, living-with-the-land repertoire of practices. Rather than import wood for huts or erect flimsy canvas tents, the women's base camp is built igloo-like into the ice, conforming with the landscape rather than imposing upon it. Instead of relying on the native bird and seal population for food, the women learn from animals and build small, hay-lined "worm holes" into the ice for bunks. In contrast to Scott's militaristic and dystopic "settle-ment," the vegetarian, egalitarian women, adapt ecologically more sound Amer-Indian survival techniques to build their underground "prairie dog village" in the "living ice" (261). An ecological imaginary frames Le Guin's major critique of the means and goals of the Scott expedition. Even though Le Guin suppresses mention of the tragedy that befalls Scott until the postscript of her story, the reader should not be surprised that his trail marks, which look so much like garbage, lead to a bad end. History, it seems, is going nowhere.

Untelling Histories

What we learn from "Sur" is that in masculine exploration history, the circumstances of discovery—who discovers the land and when and how they discover it—are more important than the discovery itself. The South American housewives place the pole in "different circum-stances." This is not to say that they see anything more than Scott did, for they admit "[n]othing of any kind marked the dreary whiteness" (270). In the women's reaction to the ultimate geographic discovery in the annals of humanity lies the problem with the whole of Antarctic exploration history: The South Pole does not precede the discoverer, or to put it another way, the South Pole is not to be discovered. It is an arbitrary point marked by no site-specific feature of any kind, a con-vergence of constructed space and time, meaningful only to those who create the artificial barriers of civilization.[14] And most emphatically, the women understand the absurdity of "leaving some kind of mark or monument, a snow cairn, a tent pole and flag." They acknowledge that "[a]nything [they] could do, anything [they] were, was insignifi-cant in that awful place" (270). Despite the difference in circumstance and the implied ridicule in leaving a phallic "pole," Le Guin has the women mimic Scott's famous cry of defiance and defeat at the pole: "Great God! This is an awful place" (*Last Voyage* 424). The women's pole too is an "awful" place. In declining to publish their exploration, the women refuse a narration of priority, echoing Scott's language without retracing his footsteps. They shift the interpretation of the "facts" of imperial exploration without offering any new deeds, or "footprints, even."

In "Sur," the material South Pole of history becomes a textual site of contention over who makes history, who writes history, and who prospers from any historical narrative. Choosing as her setting and topic the "end of the earth," Le Guin provides an alternative rendi-tion of the final moments of the regime of geographical and rational empiricism. In place of reason, progress, and linear time-keeping, Le Guin offers paradox. The feminist revision of exploration history thus makes its mark by refusing, in some sense, all heroic markings. At the same time, of course, Le Guin offers her readers a feminist utopic fantasy, belittling the engine of progress and Western imperialism by demonstrating that women have not only "been there, done that," but done it in such a way as to foil exploitation and environmental degradation. Most crucially (and problematically), the women to accomplish the task identify with (but not necessarily as) Western culture's subalterns, those already erased from the historical text of imperialism, whose silence Le Guin refashions to subversive end. While the majority of the women appear to belong to the upper classes, they are conscious of their reliance on the domestic labor of native servants to cover their temporary absences. Awareness of subaltern cultures extends to a borrowing of native Indian culture as the women take expedition names such as "La Araucana" and the "Supreme Inca." Their map names features of the terrain for South American revolutionaries such as Simon Bolivar and Manuel de Las Rosas. Furthermore, the women identify with the male crew of the Yelcho, asserting that they "were and are, by birth and upbringing, unequivocally and irrevocably, all crew" (260). Le Guin's politics of cross-identification culminates in her narrator's focus on the "backside of heroism" (261), by which she means not only the actions and agents who are not considered worthy of being historical, but also the entire endeavor to discover the materially useless polar regions. This sense of the problem of history as being shared by privileged Europeans, privileged South American colonials, and native subalterns is at the heart of Le Guin's triumphant feminist utopianism at world's end.

And yet, there is something too tidy about the women's ability to be both historically originary and transcendent. The serviceable silence of the subaltern, the corroboration of male history as the vehicle of inserting female agency, the fantasmatic erasure of women's complicit footprints—all these point to a kind of feminism that contemporary feminism has grown to struggle against. As fiction, in other words, "Sur" offers a neat solution to the problem of women and subaltern agents in history: They were always there; they simply saw no virtue in becoming "historic." This depiction in turn entails a larger critique of the search for origins, a critique which in turn implicates the very narratives by which and through which modern feminism operates. Read as a parable of feminism, the tale insists that we be satisfied with a history in which women remain marginal on one hand, and in which history continues to found itself as textual evidence on the other.

If the possibilities of feminist history seem constricted at this point, that is by design—both Le Guin's and my own. Feminism's desire to imagine itself outside forms of historical domination or, in the terms of the metaphor of "Sur," to have followed the footsteps of male history and yet to have left no garbage, creates an unresolvable contradiction. Feminism, it seems, must arrive at the same place as masculine historical narrative in order to produce its own vision of the globe, but in this, it is always belated, secondary to the "fact" of masculine origination. What "Sur" traces is feminism's utopic desire to have it both ways, to become the antecedent for male history (others have chosen to assert "pre-historical" matriarchal and goddess cultures) and to be outside of all official forms of history. In having it either or both ways, feminism finds itself unwillingly, often unconsciously, corroborating official history—perhaps especially, as it seeks to predate or revise the masculine structures of historical narrative and documentary evidence.

Read as a parable of feminist history, then, Le Guin's female expedition cannot entirely escape history, and it constitutes in itself a type of history. Although Le Guin's women leave no signs that will be recognizable to later explorers, they do leave inscriptions for one another on the ice. Utilizing the tendency of the snow and ice to preserve marks, they carve out directional and inspirational messages on handy sastrugi such as "This way out" over a big arrow pointing the way to their base. By refusing to leave footprints, the women avoid the colonialist implications of those symbols of the intrusion of culture into nature, a nature that after the impact of the footstep might be understood as having been always–already waiting for that mark—the new beginning of time. The symbols "This way out" (230) and a giant arrow represent alternatives to the footprint and all it represents. "This way out" is both a utopic gesture and a gesture to "On Exactitude in Science," Jorge Luis Borges's 1946 parable of the map of empire that grows as large as the territory itself—and disintegrates in its unwieldiness, along with empire. This fatal complicity of map and territory is complicated by feminist standpoint epistemologies and global knowledge represented by the now ubiquitous shopping mall map depicting an arrow pointing to a location with the caption "you are here." In fact, you are not "here." You are "there" in the territory that is in distinct relation to the map. Neither map nor territory is available to be fatally coordinated as Borges's modernism would insist; both map and territory are being created as "you" (another aggregated fiction) gaze

upon the map, which equally hails each of its successive viewers search-
ing for orientation, into the center of the mapped field.

What is a helpful strategy on a map in a shopping mall takes on
more layers in the context of a long tradition of imagining Antarctica
as a blank page for human action. Le Guin reprises these popular
cultural themes on the blank map of her Antarctica as a way of suggesting both the utopic possibilities of the unstoried, unwritten place,
and the real cultural limits for imagining such places. As humorous
as this moment might be, we must ask seriously: Can feminism offer
a "way out" and simply mark its own exit from the awful circumstances of history? Does the move from official narrative to journal
and personal experience somehow rescue feminism from its complicities with the disciplinary teleology of history?[15] My answer to these
and other such questions is, of course, no. The women discoverers
begin and end their actions framed by a disciplining history. Even Le
Guin's plot device, the text found hidden in the attic, is borne of the
methodology of the archive, and in particular invokes the promise
of "recovery" that formed an important period of women's history's
own history. In the retemporalization of recovery into discovery of
the lost text, "Sur" challenges not the preeminence of discovery or
archive, but the agent whose name (or gender) becomes linked to the
proof of historical ownership and thus the significance of events. In
the drive to cast women as originary agents, the feminist possibilities
of history find their limit in the priority of individualism on which
modern Western humanism depends. Le Guin's utopia of prior arrival
and of the moral, even aesthetic superiority of unnamed and commu-
nal subaltern speakers cannot hold up under the material conditions
of history, not the least of which is the complicity of white feminism
in the subordination of the so-called third world.[16] In the move from
material to discursive history referenced by the women's turning of
the blank page of history into the blank page of the undiscovered ice,
feminism implicates itself in the very histories that justified imperialism. In fact, it is the event of the women's expedition, in its renar-
rativizing of masculine exploration history, that documents the very
history it seeks to critique. A critique of the complicity with history of
an Anglo-European author taking on the subjectivities of subaltern
characters or positions might seem to most readers fairly offered yet
predictable. In the sections that follow, I complicate the critique of
"Sur"'s (and by implication, normative liberal feminism's) complicities with history by shifting the discussion from one of authorial politics to a more general consideration of the structural effects of genre
on the production of knowledge.

Undisciplining History

I have been arguing that the problematic of refusing history is figured in "Sur" by the iconicity of the South Pole, which represents both the geographical attainment of global knowledge and its epistemological limit. To the extent that "Sur" questions the value of exploration and the knowledge produced by it, I have found that its feminist revisionist history leads to a dead end, or to use a variant on Scott's description of his prior late arrival at the pole: It "forestalls" (*Last Voyage* 423) or forecloses a future for feminist and subaltern resistance. I want to argue in my conclusion that this interpretation is not only mine but the story's own. For as much as Le Guin's utopian impulse asks readers to identify with a vision of feminist history, "Sur" also undercuts its claims to a better way. To illustrate this point, I will return to the moment of geographical completion in "Sur" as the women assemble at the South Pole: "Which way?" asks one team member after the pole has been reached. "North," is the response. Le Guin's narrator continues, "It was a joke, because at that particular place there is no other direction" (270). In this way, Le Guin inscribes an awareness of the limits of the historical mode that the narrative has been both imitating and resisting. North is the direction of hegemonic power; it leads back home, both to the home of empire of the geopolitically dominant north and to the South American home of the domestic sphere. To return north leaves unchanged the axis of center/periphery, by which the south is made minor to Europe.

This problem of the superimposition of the north as originary is the subject of what Gyan Prakash calls the "impossibility" of writing subaltern history. In the aptly named journal *Nepantla: Views from South*, he writes,

> There has always been an underlying awareness that the project of "recovering" the subaltern as a full-blooded subject-agent must fail, for by definition subalternity implies a "minor" position that cannot be undone retroactively. (287)

For Prakash, the historian's desire to recover the subaltern repeats the weakness of dominant discourse by assuming that history can be made inclusive, a corrective that does not challenge history's disciplinary engine. The problem, then, lies not in the accuracy of whatever new facts historians may find about subalterns, but in their reliance on a historiographic mode of knowledge production. In place of a positivist insistence on recovery, Prakash proposes that we "understand subalternity as an abstraction used in order to identify the intractability that surfaces inside the dominant system" (288).

Such anti-disciplinary violence is shared by Le Guin. Rather than lingering on Le Guin's appropriation of the subaltern for a revisionist history, I prefer to understand "Sur" as a meditation on the disciplinary limits of historiography for feminism. This way, we avoid reproducing some of the less useful moves in the recent politics of feminism. The tale's last line, "we left no footprints, even," rather than jettisoning subalterns again from the time of history, registers in its last-minute, anxious placement a certain awareness of the problem of casting the subaltern as exterior to history and very possibly, to her own telling. Here, Le Guin is epistemologically pulling the rug out from under her carefully crafted hoax, refusing the very status her counter-history might offer. The literary allows her access to a subaltern fable that the positivist demands of historiography foreclose. This view of the construction of history as what might have/could have been saved from the indignities of disciplinary corroboration and review in real time is precisely the power of Le Guin's politics. Following those strange footprints encountered so far from the center of empire, Le Guin proposes a discontinuous past, present, and future. What Le Guin creates in her fantasy of refusing history is the possibility of creating history, of becoming an historical subject not through imitation, but through an intervention into the rational linearity of the mode of history. Like the disarticulated, "reversed" footsteps the women find on the plateau and like their own un-footnoted steps, the history of Antarctic exploration stands for the possibility of a history whose temporally distinct features of past, present, and future are discontinuous and jumbled. Le Guin's claim is not to historical veracity or justice or to any material territory. She allows a future to emerge from a past that, in the case of South American women's discovery of the South Pole, did not need to have existed in order to become meaningful in the time of history.

This is not quite equivalent to thrusting the subaltern into the role of the "outside" of history, as Prakash contends is an "impossible" position from which to create a useful counter-history. This particular outside is more precisely a before—but not the by-now familiar goddess-ghetto of archaeological prehistory. Le Guin imagines a state where those who came before, the indigenous we might say, are also understood as having their own history, as having also come from somewhere else and therefore as having the potential for a history that is outside history yet not impossible. She does not allow the indigenous to function as the natural (and thus disempowered) origin for culture; neither does the story produce an unalloyed nostalgia for

Eurocentric exploit as history. The projection of prior historical foundations has the effect of endlessly deferring origins and of shifting the disciplinary register of what constitutes evidence of a historical claim beyond the modern fact of nation or the individual and into the possibilities of confounded and cofounded origin for modernity and feminist and postcolonial thought.[17] It offers a living ice foundation for a nonmaterialist history not temporally, factually, or ideologically necessarily reliant on dominant understandings.

To be sure, Le Guin's story runs the risk of reinscribing the historical power relations she critiques through what some might consider a fetishization of a precolonial past, which once imposed guarantees the authenticity of those who can make a claim to it. But this is precisely what Le Guin does not do. Her parable against history conflates differences between British, Spanish, and US imperialism while inserting into this history protofeminist figures such as Florence Nightingale (for whom Le Guin's cosmopolitan and well-read South American explorers name a mountain) in the name of the nameless women. In this pastiche of influences, nationalism and the desire to colonize do not play a role. While a clever avoidance of specificity certainly is the foundation of "Sur"'s utopianism, the story's political force remains.

Crossing (Out) History

Bancroft and Arneson successfully completed their continental crossing in February 2001. And in their footsteps, other journeys have followed; journeys across the ice are being planned continuously. Academic writers are by definition belated, after the fact, even though we do like to imagine ourselves as originators of culture. And it is with the problem of origin and of coming after inherent in all three of my principal objects—the hoax exploration narrative "Sur," the last continent, and the anti-foundational critical discourses of feminism and the postcolonial—that I want to end this chapter. Of these three sets, it is perhaps Antarctica that retains the greatest possibilities for resisting being known as an object of human science, or for becoming fixed in history due primarily to its lack of indigenous population. It is extremely difficult to know how to ground a discourse concerning a place that has so brief and peculiar a human history and whose present and future legal and material grounding is only problematically linked to nation. Antarctica lacks the powerful ideology of origins and rights that nation provides to humanity and the territories it claims and shapes. A set of international agreements called the ATS governs the territory adequately for the present. But on the symbolic level, Antarctica remains unclaimed (if not unclaimable) and thus open to fantasies of origin, as well as fears of belatedness, of the literalization of the end of the world.

While the territory's usefulness as a symbol and setting for literature has been made manifest in part by this essay, the material acts of repetition and imitation—pole quests, crossings, circumnavigations, mountain treks, and overwinterings all undertaken in various gendered, equipped, and numbered configurations remain puzzling. I went so far as to pathologize these acts as a species of repetition compulsion. What is it that people seek in reenactments of human-geographical feats upon the vast territory? I think the answer has to do with Antarctica's lack of indigenous population. Repetition compulsion is understood as being generated by a need to confront or complete a traumatic event. Antarctica's traumatic lack of natives, while deeply epistemologically disturbing (how after all do we experience the place if not through an at least imported humanity?) is also freeing: endless successions of individuals and groups are ethically unencumbered to live out fantasies of origins and arrivals without the consequences of an imperial "coming after." Antarctica alchemically converts the leaden repertoires of nationalist and imperial territorial conquest into the gold of human striving and progress, or harmless, feel-good adventure as reflected in the contemporary fascination with Shackleton's improbable 1914–16 survival drama.[18]

But rather than seeing these acts as manifestations of a global catharsis in which a heavy human history of rapaciousness and imperial conflict is replaced by a joyful process of noninstrumental, merely symbolic feat, we might begin to fear that humanity is destined endlessly to repeat its "final" arrivals in Antarctica. In this more critical view, the terrain of Antarctica cannot sustain the overdetermination of its future. At this juncture, another future for Antarctica emerges as the ecologically devastated and plundered frozen nature preserve. The avoidance of ecological devastation as well as historical trace is the foundation on which Le Guin builds her feminist utopia. This utopic relation to the Antarctic is echoed by the 2000–1 women's crossing. But in place of literature, these women use the medium of the website to represent and to map their endeavor to a worldwide audience with little possibility of a material relation to Antarctica. And in place of native techniques of sustainable survival, the 2000 crossers equip themselves with the latest technologies of satellite communication, lightweight protective clothing, and ski and other transportation gear. Their elaborate website uses narrative, interactive journals, live web cam images, maps, and photos to advertise the women's mission and to link it to individual users. The site (www.yourexpedition.com) invites each user to identify with the remote crossers and their esoteric goal. Users can follow their route and progress as a red line extends gradually across a map of the continent. While the crossers themselves offer fairly mundane reasons for their trek (just to show the human spirit, they have always dreamed of it), the most fascinating aspect of the trek is its utter lack of utility. As I stated at the outset, the crossing was not particularly momentous as a geographic first nor as a first in adventure (solo treks being grittier). It was certainly a gendered first, perhaps even a gendered remapping of the (virtual) territory. But what type of history making is this? Making any grand claim for the feminist significance of the expedition is forced, especially given that the crossers make only the most muted and liberal reference to gender claims (as "empowerment" for girls through their vigorous example). Although they identify as American and Norwegian subjects, they make no reference to their trek as symbolic of any nation or group of people. Their website, in fact, makes no reference to polar exploration history, and does not even provide any background on previous expeditions. Any context in which the women crossers claim their firstness is curiously lacking.

The feminist utopias of "Sur" and the crossers rely, respectively, on a repression of European women's complicity in imperialism and a total refusal of all history but that of the instant web present. This latter evacuation of historical context, or rather, recontextualization of the Antarctic as a symbol (in this case for empowerment) unattached to nation or history, constructs an Antarctic that functions as a gigantic billboard for a commercialized if historically noninstrumentalized goal. The year 2000 crossers use the idea of traditional, linear history to wring value from the staging of a "first." Furthermore, by evacuating the marks of historical context through which Le Guin fashions her meaning in absence of the extra-historical "fact" of the South American women's expedition, the year 2000 crossers close down the possibilities opened by Le Guin's fable. The year 2000 crossing may in fact be the nightmare of liberal feminism, manifesting feminism's reliance on imitation and also its historical redundancy. Where Le Guin's women made history and left no footprints, the women crossers and their endless followers leave only footprints and make do without history.

Yet if we return to one of the marks left on Le Guin's textual ice, "This way out" and a giant arrow in the context of the ecofeminist concern that motivated and framed Le Guin's intervention in colonial history, a way out of the frame of history as story becomes available. The markings on ice are less important in their human-centered directions. It is the ice itself that matters.

The Limits of South South America

Pablo Neruda's 1938 invocation of Antarctica—that territory south of South America—as "There," a place indexed discursively where "all ends/And doesn't end" ("Alli termina todo/y no termina") exemplifies the doubled, perverse, and layered nature of Antarctica as a symbol of the end of the earth. Just as the second line of the poem undoes the terminus of the first, Antarctica as the symbolic end of the earth was built only to be undone. Antarctica is an end and not an end; its stones are real, terminal and symbolic; they are ever available to be indicated as terminal. Of course, Neruda as a Chilean author writing in the 1930s knows well that his direct address to the "stones" bypasses the governments of Europe and Latin America competing for a more literal direct access to those very stones.[19] Beginning with Antarctica rethinks territorial expansion of the Americas as materialist, based on stone, ice, and oil rather than word, law, and representation. As economic and climate disaster, ecological politics and postcolonial conditions increasingly intersect, interrupt each other, they complicate the terms of formerly singular cultural resistances such as feminism. "Peak oil" creates new desperate modes of territorial claims through the sea beds and attempts to produce Antarctic ice as a "commodity frontier." Countering the dominant legal reformation of the Southern Oceans under European-centered treaties, this section takes up ice as unaccountable within fiction, poetry, film, and performance at the material, environmental limits of Antarctica.

What environmentalists call "hard limits" of a resource-extraction civilization combine discursive structurings with materialist ones. The fantasies and fears around climate change and survival connect the Americas to the modes of projection onto Antarctica that render it unlivable yet in need of conquest; a shared international territory that is closed to anything but an incompletely demilitarized scientific management. Seeing Antarctica within a new materialist frame complicates existing anthropocentric models of postcoloniality, globalization, and population construction, particularly in relation to the geo-reconceptualization of Antarctica as "South South America" initiated by Ursula Le Guin's "Sur." To recap from the previous section, "Sur" presupposes a pan–Latin American women's expedition to reach the South Pole. Beating the official heroes Amundsen and

Scott to the pole, the women admit that they "had no business being there," and to further undermine imperial constructions, leave no footprint or trace of their claim. The way Le Guin both closes and leaves open this geo-imperial history loops into a set of nesting discursive and material ecologies for understanding what "South South America" might be in a global ecology characterized by constriction, climate disaster, and material limits of resources, especially of water and oil.

Antarctica is not necessarily extensive with life—bio or geo—or with politics—post or neo-. States, individuals, corporations, affective flows, and medias converge and disperse across the ice in blocks of intensity that cannot be understood in conventional formation, or even as postnational or transnational. "South South America" is only one way to speak of the extensions and disruptions offered and repelled by Antarctic materiality, by its history, and by its refusal of the kind of history instigated by Le Guin's insertion of South South America into the obliterating historical container of Anglo-European global exploration. Echoing Neruda's expansion of terminus and the termination of expansion itself, Le Guin tries to balance the critique of Euro-historical geo-power with an ecological sympathy for material earthly limits that postcolonial resistance cannot automatically or necessarily undo. When her counter-explorers reach the South Pole, they joke about the arduous arrival, asking now "Which way?" The logical response, "North" speaks to the imaginary limits of movements of resistance, to be sure. But it also points to material limits subtending politics and imaginaries that we now can ever more directly label environmental–economic disaster under oil capitalism.

Given the framework of a materialist study of the earth, what is "South South America" beyond a naturalized extension of European imperialism? There are no clear victim–actor negotiations for Antarctica. Despite the discursive politics around firsts, arrivals, and feminist and postcolonial revisionism, no natives have ever been harmed or oppressed on ice. Which is not to say Antarctica is, as the cliches insist, a "continent of peace" or an exception to the global history. Imperial history has left more than symbolic, ephemeral marks across the ice; the heroic dead white bodies littering Antarctica's human history are no expiation for European imperialism, especially as that history continues in renewed form to leave its mark.

Neruda and Le Guin's undocumentable Latin American Antarctic histories that paradoxically both imprint and self-erase are, more creditably, affective possibilities for South South America that counter even South American postcolonial culture in Antarctica. Rather than press recursive or depleting claims, artists continue to work with the affective possibilities and limits of Antarctic materiality. For example, Belgian-born Francis Alÿs's 1997 performance and video "Sometimes Making Something Leads to Nothing" ("Algunas Veces el hacer algo/no lleva a nada") consists of the artist pushing a block of ice through the streets of Mexico City until it completely melted away. The implications of this process performance resonate ever more in the context of Chile's 1992 display of a towed and reassembled iceberg at the Seville World's Fair, and the attempts of corporations to bottle Antarctica's ice resources for global distribution (Brazil has a water brand "Antarctica"). Alÿs's paradoxical creation-by-depletion mimics the sacrificial expeditions of the dead and wounded European explorers (also praised and mocked by Le Guin), inviting a "reading" of its trail of ephemeral water-ice. Labor and artifact loop and consume each other as they are also self-consumed. The search for fresh resources and the trail of migrating (or towed) ice and human labor is neither strictly legible nor at all triumphal; it is more like a diminishing and finally disappeared Sisyphean task: more frightening for the neoliberal imagination than the doom of endless, pointless labor (think of all those heroic crossings and treks) is the idea that doom itself might come to an end.

Alÿs's melted ice points away from the limits of capitalized environmentalist strategies such as recycling, and instead to a new "Hole at the Pole" or an end that is not an end. For the near future, polar wastes will be redistributed not necessarily as material but as affective flow, as melancholic attachment, as paradoxical process: in short, as globalization in ruins. Alÿs's intense and pointless labor demonstrates such tortured attachments. Art (and heroic polar exploration) is affective labor, especially one that seeks suffering and perversity, speeding needlessly to an end that is inevitable: dissolution, loss, death. The only way to avoid loss is, after another one of Le Guin's postcolonial feminist koans in "Sur," "to carve in water." More specifically, the "penalty for carving in water" (265) refers to the paradoxical labor of one of the Latin American women expeditioners of the story, who creates a series of sculptures using the only available material: ice. Her sculptures, however, cannot become marketable and must remain forever in place. The "penalty for carving in water" is another turn on the problem of all extractive, colonialist exchanges, even those with the nonbiological material earth. To become one with the territory and time of ice requires forgoing signification in official history and value within markets of exchange. And since that is equally impossible (or unattractive) for humans and for resistance movements as well as dominant culture as presently configured—all we have are these glimpses of the nothingness, the cold, the non-entity that we might become on the other side of the limit. Strict ecological materialism begins to seem too much like pushing an iceberg around until it melts away.

These limits might translate to the end of territory for colonization and thus to the redistributions and recuperations that characterize postcoloniality and the new ecological materialisms it informs. Yet, this hard environmental limit, just as the geographical limit of the South Pole, was virtualized from its inception. Edgar Allan Poe's 1838 nightmare of south polar exploration, *The Narrative of Arthur Gordon Pym*, is a good place to begin a new history of reinvigorations of the end of the earth that resist or at least try to resist dominant capitalizations and instrumentalizations. There is no underestimating the growing importance of Edgar Allan Poe's seemingly benign picaresque of a boat voyage from New Bedford, Massachusetts, to the ends of the mapped globe. Its final vision of the polar abyss coincides with the end of the story, a drive to the whiteness at the end of the page that nicely wraps up a history of Eurocentric discursive constructions of Antarctica—a form of reading that this materialist, postcolonial analysis pressures. Subsequent revisers and completers of *Pym*'s tale of the end of the earth have also come up against those discursive and hard limits at and of the South Pole. By 1897, Jules Verne explains Poe's abyss as a giant lodestone inexorably pulling in ships, a technoscientific horror plot enveloping human time and navigation. H. P. Lovecraft, Poe's most assiduous imitator, offers a similarly teasing end of endings, that like Poe's used the narratives of contemporaneous expeditions to substantiate and undermine official accounts of the Antarctic. *At The Mountains of Madness* (1936) exploits US Antarctic explorer Richard Byrd's language describing a land "beyond the pole" and the many "lost race" fantasies set in the Antarctic to invent a scientific expedition that discovers the hideous true creators of the human race dormant under the ice. Ignorantly penetrating the polar abyss, the scientific team is ejected from the ice itself by a volcanic eruption. Lovecraft's proleptic apocalypse through the extrascientific discovery of what is the hopelessly inferior and belated position of humans and scientific knowing incarnates the problem of hard limits as one of (repressed or unknown) origins as well.

The hopeless position of humanity within the ecology that science has developed to describe and to study is a modern problematic of almost endless productivity. In other words, the end of the earth is a perfect plot. Not only have reimaginings of this discursive end of the world at the South Pole persisted, they have looped with traditional film and new media. *The Truman Show* (1998) visualizes the perceptual world of the mass-mediated human imagination. The protagonist, caught in a reality show that he only slowly becomes aware of *as* his own life, attempts an escape from—or at least an investigation of—his condition. Gesturing back to Poe's and Verne's nautical explorations, he takes a boat to seek the edge of the world. Unlike the children of *Pym*, Truman does not fall into the hole at the pole. Instead he bumps up into the hard limit of his mediated earth. An invisible dome encloses his perceptual universe. What naturalist Jacob von Uexkull writing in the 1930s called a creature's "umwelt" applies to our mediated umwelt: that "bubble of perception, or the lens in, of, and around our mediated beings." But, as these literary and filmic virtualizations infer, a limit that is also a lens or a text (or another form of abyss) complicates a natural historical account of human–other interface. The ice itself, like the blank page or the camera lens, can be one of those paradoxical abyssal hard limits. Ice might be termed a mediated hard limit.

I discuss this 1998 film about US-mediated culture because that theme of discursive, perceptual limit replayed from geo-metaphysical literary precursors set at the South Pole has been itself retaken up in contemporary Antarctic fictions. The 2006 animated film *Happy Feet* sends its dancing penguin solo to the end of his world. He is seeking an answer to disastrous changes in his umwelt. What he discovers, much like Truman, is that his world is limited; what is outside it is near impossible for him to conceive of, much less to translate back to his own people. His story of huge ships and pale people has less of a chance of being believed than the speculations of the colony's wise man, who has refunctioned the plastic soda can holder that is slowly strangling him into a talisman of his otherworldly power. Eventually, the little penguin is captured (or as environmentalists would have it, rescued) and put inside a zoo where he perceives himself become the perceived. We have all stared at diving penguins through thick glass aquarium walls. This glass containing the world and not-world, the human and other, the living and dead has never held. It has always leaked. Antarctica—with or without its native animals—has become the most available earthly location for the actualization of perceptual and now hard limits as they collide in nesting yet incommensurable ecologies of perception, affect, sensation, and being.

Grafting European and African worldviews of multiple and nesting ecologies of life and non-life, Kevin Brockmeier's 2007 novel *The Brief History of the Dead* imagines the end of the earth as a result of global terror, rampant corporatism, and ecological collapse: in a near future of warming and extinction, a virus wipes out humanity even as a US-multinational soft drink corporation sees this disaster as an opportunity to market the last pure water remaining on earth—in Antarctica. Like its precursors *Pym* and *The Truman Show*, the novel is a tightly nested ecosystem of preface, chapters, and inconclusive conclusions. Its two universes—familiar earth and the City—live symbiotically in alternating chapters. Following the African belief system, the population of the city of the dead rises as people on earth die; but they only remain in the city for as long as people on earth are alive to remember them. As the virus creates mass human extinction, the City also empties of its forgotten dead. The residents of the City search for answers to their sudden depopulation by exploring the City's outskirts. There they run into a recursive and hard limit, again, a sort of a dome: the edge of their umwelt. Parallel to this narrative is the story of a mid-level Coca-Cola employee named Laura Byrd (a reference to Richard Byrd, the once-famous US Antarctic explorer of the 1920s and 30s) whose publicity stunt trek across Antarctica has gone wrong. Surpassing her namesake, who nearly died attempting to overwinter solo, Laura Byrd achieves the ultimate sacrificial heroism slowly starving to death alone in a tent on a vast glacier, isolated from the virus that has wiped out humanity—the only and last person alive. Her memories—and thus the inhabitants of the City—die with her. As Poe's serialized and failed novella gestured self-knowingly to the limits of literary and market-driven production with the blankness of the end of the page, the hard limit of extinction and biological-climate disaster connected through capital rapaciousness, are both echoed and contained through the affective limits of memory.

Of course, all these limits—blankness at the end of the page, polar abyss, melting ice, cultural amnesia, literary market failure, and biological extinction—are also material, hard limits. This discursive limit to our umwelt is also a hard limit. As it turns out, Alys's iceberg did not need towing. The ice is migrating through transnational capital's desperate search for oil, as a polar condition: The entire earth is "becoming polar." As I elaborate in the epilogue, events such as the 2010 British Petroleum spill in the Gulf of Mexico demonstrates this polarization of the entire earth as wasted, available (for drilling), and in desperate need of rescue. As polar limits are redistributed all over the map, a new relationship between the discursive and the material must be imagined, or created. To return to Le Guin's terminal question: Which Way? Can postcolonial feminism focused on South South America, if not find a way out (there is none) at least reinvigorate the problem of territory/biopolitics in postcolonial theory?

In an earlier era of postcolonial discursive and material claims on Antarctic territory by Chile and Argentina, Pablo Neruda understood the anxiety of the end of the earth: "there all ends/and doesn't end" as the engine of postcolonial competition. Antarctica as an imagined end of empire, or as "South South America" is the end that could not be the end—an impossible ending that dovetails with Le Guin's postcolonial koans hidden like word-bombs throughout the text of "Sur." Yet, Neruda's opening of the south and his approach to its materiality of stones is not entirely compatible or complicit with the geoeconomics of postcoloniality. Neruda's 1934 poem "Stones of the Sky" upturns and involutes temporal and environmental ecologies.

It would be a mistake to read "Stones of the Sky" as a version of South South America of postcolonial extension. Although the poem begins with a classical lament to the forces of nature to "wait up" for the presumably human speaker, Neruda does not linger on this sense of lag between human and nonhuman time frames. Rather, the poem advances through nondirectional and indefinite tone, flows, dustcloud, and lava. Each of these phases, which appears in the poem as verbs, abstract nouns, and concrete things flow together to a "point, or port, or birth." This birth, or origin point is not of the human, despite the possessive sense of the final word, "ours." What is for humans is for the matter of the earth and the universe of stars: The stones of Antarctica flow into the "snow/or caked dust in the desert/regions, metallic/dustcloud." What is this metallic/dustcloud but Antarctic stones, Antarctic "icy sapphires" sublimating, melting, and dispersing through the air, to be breathed in by all animals and plants as well as crystal structures comprising ice and stone? "We shall be stone": if becoming stone flows into "borderless night," Neruda's refusal of borders—or his de-linking humanity's umwelt from the engine of survival as reproduction—takes specific, obliterative, shape in relation to the force of postcolonial–neoliberal international science management, for which global climate disaster becoming productive through and because of the hard limits of ice and oil.

Chapter 3

"Who Goes There?": Science, Fiction, and US National Belonging in Antarctica

Imagining a future in Antarctica has never been easy. British explorer James Cook on his third voyage around the globe in 1775, frustrated by the frozen seas, the absence of arable land, mineral wealth, or anything he saw as valuable, declared the imperial territorial quest ended: "no man will ever venture further than I have done...the lands which may lie to the south will never be explored."[1] Cook's impulse was based on his extensive experience in the region. Indeed, the preceding centuries of Antarctic exploration had seemed more like un-exploration in that every new mark south erased a fantasy of a warm, habitable, and most importantly, profitable region to the south. Seeking to extricate himself from an increasingly bad investment, Cook staked the bottom of the map, overwriting the Greeks' previous imaginary, *terra australis incognita* (the unknown southern land), with his own terminus of impassable ice.[2] Cook's mark stood for barely 40 years before a frenzy of sealing voyages gave way, once the seal population was dispatched, to further territorial conquest. His instinct to make Antarctica the limit of modernity's drive to know the world is echoed in the persistent need to place a future—a concept that insists upon some sort of value to be produced—in Antarctica, a future that Antarctica's materiality thwarts.

Antarctica's remoteness, lack of indigenes, and almost otherworldly harshness has prevented the development of even the crudest industries, save for tourism. While tourism continues to grow and reward a range of private and self-regulating companies—exploiting for profit the very pristinity its industry threatens—my interest is in

Antarctica's resistance to the types of resource extraction typically pursued by nations either within their own borders or under colonialism. In its vast and mostly desert-like qualities, Antarctica challenges the way capital has mapped the earth into zones of productivity, themselves resolved into further layers of industry or labor-value.[3] For a long time, Antarctica was for all intents and purposes what Cook predicted: a wasteland of marginal human concern, resistant to all but a few hardy and well-funded expeditions.[4]

But Antarctica's salience has risen since WWII, especially as a result of the development of the 1959 ATS and the science activities occurring under the IGY. The ATS sets Antarctica aside from competition over territory and from capital development, instead designating it for the sole benefit of international science. Standing firm as a legal structure for approaching Antarctica, the ATS and its member states have through the 1970s and 1980s added amendments and protocols to address renewed interest in resources, protect its environment, and consider inclusion of nations with relatively little historical connection to the continent.[5] The 1991 "Pax Antarctica," or the agreement to extend earlier deferrals of contests over territory, power, and policy, re-ratified the basic tenets of the 1961 ATS, banning militarization and commercial resource mining and deferring national claims until 2048. The implications of the ATS regime have been profound, both structuring the possibilities for states to engage Antarctica, while limiting those very engagements to those directly related to science. State-run science has in many ways solved the problem of Antarctica's troublesome materiality and provided a safe course for national rivalry. Yet, as international investment in Antarctic science continues to grow and human presence with it, with the United States leading in expenditures and personnel, even powerful international science and the ATS cannot prevent the simultaneous development of territorial and other types of property and sovereignty-driven interest in Antarctica.[6]

The era between the world wars was a key time for the development of US national interest in the region. The United States finally had in Richard Evelyn Byrd (1888–1957) its first major government-funded expedition since the one that had inspired Poe, the US Exploring Expedition led by Lt. Charles Wilkes in 1828–32. Byrd extended the Heroic Age's emphasis on crowd-pleasing geographic firsts with his 1928 flight over the South Pole into the more technological and science-driven modern era marked by the ATS and the international science of IGY. Byrd's exploits made him famous in the United States, and his memoirs, documentary films, and journalism worked to sustain attention to Antarctica and to inspire political support as well as popular response, some of it in the form of fictions influenced by his published narratives.

The most significant depiction of Antarctic science written in the era before the ATS took effect is the 1938 short story, "Who Goes There?" by John Campbell, Jr. Set in a research camp in the middle of Antarctica, the story tells of an alien discovery with the potential to wipe out the human race. The alien attacks its discoverers, those who would study it, and the story becomes a philosophical drama about the methods and ethics of scientific study, and of the dynamics of polar colonization within an isolated homosocial group. Tracing this story of the threatening alien "thing" through its two filmic iterations—*The Thing from Another World* (1951) and *The Thing* (1982)—this chapter examines the implicit contention of contemporary governance in Antarctica: that enlightened science under national programs is the best way to secure Antarctica as a place of value. The chapter therefore concerns itself in part with the ATS as governance structure and specifically with US national science, arguing that national presence in Antarctica is not benign, natural, or necessary. It also challenges the consensus celebration of the resiliency of the ATS, and instead questions the forms of human activity and inhabitation it has allowed, and considers the possibilities of governance it has defended against or even permanently forestalled.[7] Antarctica is thus a place more complex and layered than its popular namings of "white desert," "continent for peace," or "frozen laboratory" suggest.

Barely registering within a history of European Empire, Antarctica has most often been placed within British Edwardian cultural history as an extension of British imperialism—but an extension that provides a relief from that very history of unequal conquest—a clean white space, where Europeans could pursue an "unashamed heroism."[8] The race between Norway and Britain for the South Pole seemed to sublimate imperial conquest to an elemental struggle to survive. Yet once attained, the South Pole lost much of its topicality, or its aura as a plot generator. While fascination with survival and all it implies about the linked fragility of masculinity and nation persists to this day in both serious and parodic retellings of the Heroic Age of Antarctic exploration, Antarctica's shift from a purely symbolic prize to a region connected to global processes of capital and human culture is only recently underway.[9]

Lacking direct connection to the Heroic Age of Antarctic exploration of the turn of the twentieth century, US study of Antarctica is even less developed. Limited public cultural knowledge of the region, the low political salience of the territory, and American exceptionalism, or belief that US history breaks with that of Europe, has resulted in what amounts to a near universal lack of Antarctic knowledge in the United States that is only recently being addressed.[10] The study of Antarctica has been doubly distorted in the United States by exceptionalism that refuses to see US interest in Antarctica as connected to discourses of empire, and within American Studies itself, in which the region has only recently figured as a space or object of study. Without positioning Antarctica as "outside" American Studies and thus running the risk of creating an exceptional object in the course of critiquing exceptionalism, it is nevertheless useful to see Antarctica as outside conventional limits of national studies. Approaching Antarctica as outside convenes recent turns to hemispheric models that seek to break up naturalized spatializations such as north–south, east–west, and on transnational or postnational conceptualizations attending to flows across borders of historical periodization as well as of geography.[11] Following recent discussions of Transnational American Studies on foundational repressions and assumptions of empire in US national formation and international relations, this chapter suggests that US empire is distinct in Antarctica: not absent, but different and understudied, in an area both conceptually and geopolitically "unsettled."

European and Commonwealth nations after Cook struggled to integrate Antarctica into their territorial and capital expansions. Britain, Norway, Australia, Argentina, Chile, and New Zealand have all anchored their interests through territorial claim.[12] However, the United States (and the USSR) has never registered claims, though it retains the option of future claims. Yet, today, the United States functionally occupies the South Pole with its new Amundsen–Scott Station and maintains the most lavish and well-funded bases in the continent, anchored by the largest base, McMurdo Station.[13] The tension between territorial claim and other forms of occupying Antarctica—in particular science—is displayed by Byrd's career in which contradictory forces within the United States vied to solve the problem of what to do with the enormous fact of Antarctica.

Thwarted Hero

Like Cook, Byrd was thwarted in attempting normative territorial empire in Antarctica. But unlike Cook, it was not the impracticalities of ice that thwarted him, but rather the US refusal to claim the territory Byrd had brought into the gaze of the nation with his 1929 overflight. A Navy flyer, Byrd had hoped to be the first to cross the Atlantic. But Lindbergh accomplished that feat in 1928, and Byrd immediately turned to another flying route bound to reshape the earth: the path over the South Pole. Although a remote location outside the circuits of civilization, the South Pole nevertheless suggested a symbolic achievement and a prize of technological mastery. Flying over it for the first time augmented symbolic US international power, yet the remoteness and disembodied nature of aerial reconnaissance stopped short of transforming Antarctica into a US national territory. Nevertheless, the feat made Byrd a hero. Unlike the flight across the Atlantic, which symbolized a shrinking world, speed, and the ability of the US to exploit connection with a formerly central Europe, Byrd's flight to the South Pole awakened frontier fantasies in the United States. Would the South Pole become a new center of an internationalizing world, or an extension of US territory? Byrd would spend the remainder of his career—spanning the Depression, WWII, and the beginnings of the Cold War and the IGY—on the question of what this territory might mean to the United States.

Byrd's Antarctic career breaks down to three distinct phases. The first of these is the South Pole overflight of 1929 and his subsequent expedition to establish the unsubtly nationalistically named "Little America," an encampment on the Ross Sea ice that would have five iterations and lead directly to the establishment of a permanently manned station at the South Pole in 1961. In 1934, in the midst of the expedition that would precede the series establishing "Little America" base camps, Byrd solo wintered at Advance Base, a dangerous and near-fatal personal choice (many considered it a stunt) that resulted in even greater fame after the appearance of *Alone* (1938), his memoir of the expedition. The second phase of Byrd's career began in 1945 at the conclusion of WWII, when he urged an immediate expedition, arguing that the excess military buildup from the now-finished war would provide men and equipment for a US Antarctic claim (Rose 2008: 427). Byrd's high status in Washington through the 1940s, however, diminished under Truman's presidency. In this final phase of his career, Byrd lost the command of the Navy's Operation High Jump expedition to George Dufek. Yet, throughout his career and despite the decline of his influence, Byrd maintained promotion of Antarctic internationalism as a key to a "new world peace structure" (Rose 2008: 426).

Many cite Byrd as a major architect of the IGY, and thus of the ATS.[14] While such accounts of Byrd's influence certainly pay tribute to his sustained contributions, others discuss the tensions emerging among the State Department, the office of the President, the Navy, and the Congress over the US presence in Antarctica and specifically how they should adjudicate Antarctic claims, including the possibility of an official US claim (Carter 1979). The tension between different modes of presence in Antarctica—should science take the lead? Should industry be developed? Should military uses remain central?—were played out in Byrd's attempts to solve the problem of Antarctica, and were reflected in his writings and the popular press reports on his efforts. The difficulty of translating Antarctic ice and scale into a US imaginary of territory in the years after Byrd's overflight and up to the IGY of 1957–8 is evident in coverage of his Antarctic exploits in such popular US media outlets as *The Saturday Evening Post*, *Reader's Digest*, and *National Geographic*. Despite the lavishly illustrated spreads declaring the Antarctic a "winter wonderland," and extolling the can-do efforts of the hardy colonizers of the sublime landscapes, no amount of daring action on the ice could satisfy the ultimate demand for useful territory and locations for capitalization. In interviews and articles, and in his own writings, Byrd worked hard to counter the persistent question, Why Antarctica? To Byrd, the value of a US claim in Antarctica seemed obvious. Throughout his career, he produced papers suggesting the economic value of mineral deposits, the importance of Antarctica as a training site for the US military, its strategic significance, as well as its centrality to scientific study. Nevertheless, many of his 1930s' publications and recordings strain credulity. Would the untapped potential of free ice-space invent Antarctica as "the icebox for the world" and thus solve the "terrible problem of stale bread"? Even to the most patriotic mindset, the idea of Antarctica as a "perpetual granary" must have sounded desperate.[15] Post-WWII US empire had simply not developed to accommodate an occupied Antarctica.

Byrd's more straightforward hopes of claiming Antarctic territory ran aground on the influence of the 1924 Hughes Doctrine in which the then secretary of state declared only permanent colonization of territory adequate to constitute territorial claim.[16] This meant that Byrd's marker-dropping from the air and other symbolic acts could not anchor claims.[17] The Hughes Doctrine was undoubtedly a spur to the colonization inherent in the "Little America" installations. Yet, despite considerable congressional and public support for extending a US claim to Antarctic territory, what precisely might count as adequate colonization was never spelled out and neither Little America nor any subsequent US action in the region has stood as a basis for a claim. Marked by the official claim that never came, Byrd's career reflects the incoherent, multiple approaches to making Antarctica sensible as a US territory (Templeton 2000). Byrd was left to counter with a new imaginary—one that he had actually been developing

since recovering from his 1934 near-fatal solo wintering at Advanced
Base in a hut that was if nothing else a transparent gesture to classic
frontier-style colonization.[18]

In *Alone* (1938), his account of his wintering, Byrd sidestepped
the issue of territorial expansion through colonization by emphasizing the scientific value of his weather data collection and by directing
attention to his personal struggle to survive carbon monoxide poisoning from a faulty exhaust pipe. Even as he describes the Antarctic
in familiar terms as being vast and empty and full of danger, his
mission, that he admits is negligible "aside from the meteorologi-
cal and auroral work," removes Antarctica from its associations as
proto-national territory and repositions it as a mythical location for
personal stocktaking. Much of the tale concerns domestic, day-to-
day activities around the hut: Byrd's routines for eating and keeping
warm, the terrible pain of his poisoning, and the havoc that wreaked
on his sanity-preserving routine. By taking time off from what he
self-consciously referred to as the "hero business" to become a mere
caretaker of weather equipment in a humble hut, Byrd subverted
heroic narratives of Antarctic occupation. Yet, his narrative was easily
recuperated into an American romantic vision of Antarctica by author
and Antarctic biologist David G. Campbell, who sees Byrd's Advance
Base as a version of Walden Pond, or a landscape connected to Byrd's
boyhood in Virginia in which he relived a childhood "feeling of peace
and exhilaration."[19] *Alone* signals a mid-career vision of Antarctica
in which Byrd makes complicated arguments for the instrumentality
of the region for nation and begins a campaign to reassign meaning
to the ice through personal memory and a connection to iconic US
landscapes. As Stephen Pyne notes, Byrd developed a frankly spiritual
vision of global brotherhood and peace at the South Pole.[20] Late in
his career, Byrd expressed his dreams for Antarctica's future based on
his memories of Antarctic terrain:

> The vastness, clearness, whiteness, silence, the purity, the elevation
> above the petty quarrels and ambitions of men and nations, combine
> to form a majestic symbol of what man should want most, peace on
> earth.[21]

Byrd concludes with a separate one-sentence paragraph: "Antarctica
is a sermon in ice." In this symbolic recasting of the territory, the
Pole is no longer a remote and extreme landscape absent of human
meaning, but rather a new center space of the spirit of disinterested,
territoriality-free globalism. No longer the answer to the "terrible

problem of stale bread," the Pole occupies a mobile symbolic center of global realignment for peace. Frustration over the intractable qualities of Antarctic materiality, coupled with the unchanging US policy against claimancy drove Byrd to invest Antarctic ice with a vision of global peace "located" at the South Pole.

Byrd's irenic flights may have been as foundational as his overflights in offering a distinct and much-needed direction away from the international competition that had characterized culture at the Pole up to that era. However, these internationalist visions emerged from Byrd's continuing struggle to assert a more nationalist and instrumental presence in Antarctica. Therefore, the paragraphs leading up to the vision of peace at the Pole enumerate Byrd's by now well-established arguments in favor of Antarctica's value to the United States as a source of natural resource for a nation "squandering" useful minerals, and as a "proving ground" for the military. How can we account for Byrd's two futures—one in which Antarctica would become occupied and mined for the needs of the United States—and the other in which Antarctica served as a location of international peace? The disparate discourses of science would negotiate Byrd's contradictory approach to Antarctica.

Byrd's legacy leading up to the IGY and ATS needs to be more carefully articulated, not only for what it can tell us about US imaginaries in Antarctica, but for its effects on contemporary Antarctic science-based culture. After the publication of *Alone* and through witnessing the destructive potential of technology in WWII, Byrd became increasingly suspicious of science's ambiguous applications and unstable effects, or its ability to "prepare a cataclysm which will bring to final ruin all we have achieved in the last three hundred years" (qtd. in Rose 2008: 390). While it would not be far off the mark to ascribe this quote to Byrd's personal frustration about the ruin of his Antarctic career's goal to claim territory outright, what is most significant is his ability to question how science functions both to ground and to unsettle his dreams. This problem also links Byrd to the ongoing complexities in the ATS's codification of science as the sole sanctioned activity in Antarctica.

The ATS respatializes Antarctica; that is, its development has put into play an Antarctica of territorial possibility by deferring competing claims into the future and by instating science—the quintessential method of knowing, and basis for capital development—as the sole sanctioned activity. Yet, the ATS represses as many possible Antarctic futures and spaces as it allows. And more importantly, as the effect of the replacement of capital development by the ideal of pure, international science continues to be expressed in proliferating national science programs and their infrastructures and personnel, the ATS no longer describes the territory it nominally regulates. Recent developments in Antarctic geopolitics demonstrate the limits of the ATS to describe the future of the territory. On October 17, 2007, Britain, a founding nation of the ATS, activated its three major territorial claims in Antarctica. It did so using the Law of the Seas (LOS) to claim strategic seabeds and thus potentially valuable areas of oil reserves, which is in clear defiance of the ATS, which states that no claims can be acted upon under its regime. Ironically, the ATS is cited as the major model for the development of the LOS, illustrating one of the ways that embedded in the ATS is its own demise. While the British claim has little to do with science (and thus all the more sidesteps the ATS), it nevertheless pressures the regime so that the question must be raised: What alternative futures have been obscured by or made impossible by the success of the ATS and its promotion of science as sole mode of human being in Antarctica (Salleh 2008)?

Thawing Out the Thing(s)

Science is government in Antarctica: It is a *raison d'être*, alibi, material condition, de facto governance structure, and epistemology.[22] Since IGY and ATS, it has taken up the burden of human presence and dominated visions of the future for Antarctica. This section discusses the implications of science becoming the hegemonic mode of engagement in a vast territory in which the traditional forces of nation and capital markets have been put in abeyance.[23]

John Campbell Jr.'s 1938 "Who Goes There?" (written under the pseudonym Don A. Stuart) came out in an era of US turn to domestic, anticolonial policies. The New Deal attended to the damage inflicted by markets failing. Isolationism became increasingly attractive and would only be quieted by the United States' late entry into WWII. Antarctic policy, governed by the restrictions of the Hughes Doctrine, was very much in line with US antiexpansionist policies. Scientific–military ventures such as Byrd's were thus caught in a bureaucratic no-man's land, even as the plot of a remote outpost at the South Pole captured the popular imagination through Byrd's exploits. "Who Goes There?" centers on a scientific expedition's discovery of an alien frozen in the ice, and the scientists' attempt to understand, battle, and ultimately destroy what they refer to as the "Thing." The Thing has the ability to reproduce itself by imitating all life forms it contacts, individual by individual, until no originals remain and all life becomes the Thing. It can control dreams and read minds of the individuals around it. It conquers through incorporation and mind control, not through physical confrontation. Its effects are detected through the breakdown of morale and symptoms of paranoia and other forms of mental disturbance, and baleful looks between the men. Although originating in the embodied discovery of an alien, the threat of the Thing extends beyond the borders of bodies and into the realm of ideology.

In the story, the scientists argue over whether to study or destroy their discovery. Blair, the biologist given to effeminate "bird-like" hand motions, has an almost unseemly desire to study and it is implied, connect with, the alien. The hero MacReady is a meteorologist who has no doubt that the Thing is a dangerous and "evil" entity that must be killed, not studied.[24] MacReady is massive and bronzed, a monument of masculinity and surety. His association with precious metal hints at a value to scientific presence beyond the research that ostensibly motivates the discovery. But what is discussed in detail is what to do with the discovery of a seemingly dead alien. The alien's tentacles and malevolent red-eyed glare inspire fear and disgust in its discoverers. While the scientists debate holistic relativism versus a rigidly human-centered approach to knowledge, the plot settles the argument on the side of absolutism as the biologist's "pet" is accidentally thawed and begins its assault. The alien's method of assault is to infiltrate the minds of the men and then take over their bodies by imitating them, man by man, until no originals remain. The epiphenomena of this process of reproduction include paranoia, as the men no longer know who is real or an imitation. Their inability to discern dream from reality results in an ontological horror in being imitated, or of becoming not themselves. Empiricism, however, comes to the rescue as MacReady devises a "blood test" to distinguish real from imitation man. The hard scientist quells the existential crisis caused by the biologist's lust for dangerous knowledge. The alien is detected and vanquished—though this requires that the "real" men chop to pieces and incinerate the "imitations" among them, who look and act exactly like the real men. The extirpation of "Things" who are indistinguishable from the men reestablishes the order of species as well as the authority of science over questions of the human even as it points up the instability and paradox of humanism.

Perhaps the most under-discussed aspect of the Thing narrative is its powerful spatialization of Antarctic territory.[25] "Who Goes There?" is a story of contested colonization and overlapping claims in the guise of an alien takeover narrative. As the scientists deduce from the evidence they discover buried in ice, the alien crashed into earth 10 million years ago: It is both colonist and rival explorer; most importantly, it is a temporally prior inhabitant of earth. The human sojourners note that the Thing, like them, has come to Antarctica from outside its boundaries. But rather than employing scientific objectivity to begin to account for the conflict over Antarctic territory or its multiple inhabitation, the prevailing ethos of the tale is one of defense and counterclaim (although the Thing's method of aggressive assimilation does not leave much room for negotiation). Campbell's narrative evokes an Antarctica of spatially and temporally overlapping claims. That the Thing's revival takes place through a thawing—the characteristic marker of temporal or spatial change in ice—underscores the connection between the Thing and the shifting, unsettling place of Antarctica, an association also reinforced by the grammatically shapeless and protean noun thing.

The story's war game veneer and noisy flame-thrower battle scenes between the men and the Thing hint not very subtly at a political context of brewing international contestation in Antarctica and polar territories more generally. The 1938 Nazi expedition to found New Schwabenland in Antarctica by dropping aerial markers from a plane so alarmed President Roosevelt that an official US Expedition went into planning, and Secretary of State Hull issued a statement reiterating the gist of the Hughes Doctrine against recognizing territorial claims as they arose from Argentina and Norway. "Who Goes There?" suggests the limits of both science and of colonization, and thus might seem to support the Hughes Doctrine's stance that only long-term colonization by the state formed a valid basis for territorial claims. But in its reliance on uninterrogated notions of science to reestablish a sense of order, the tale of the Thing undermines its own logic, and thus the spatial controlling of Antarctica as remote, cleansed location for the pursuit of scientific knowledge under the aegis of nation.

Frozen Laboratory

Science became increasingly the center of US military-run expeditions in Antarctica in the years after WWII up to 1957, as the IGY instantiated the shift from geopolitically open to scientifically purposeful occupation. Byrd's Little America expedition resulted in no permanent bases. Beginning in 1947, Operation High Jump sent unprecedented numbers of men to live in Antarctica. That its establishment of permanent bases still did not provide the rationale for an official claim reflects the disconcerted policies toward Antarctica in the postwar years. As Jason Kendall Moore has argued, the US anticlaim position (which persists into the present) also reflected the greater geopolitical and practical salience of the Arctic. The emerging Cold War and the increasing role of science as the preeminent and politically instrumental mode of engagement in Antarctica account for the reemergence of the Thing narrative and for its relocation to the North Pole in Nyby and Hawks's 1951 *The Thing from Another World*. In the opening scene, bored officers playing cards remark that the "Russians are crawling over the pole like flies" to establish anxiety about a post-WWII US military grown lax, and of tension between the methods and goals of science and those of national security. Rather than dividing science into competing subdisciplines, the 1951 film distinguishes crudely between scientific and nonscientific cultures. Carrington, the sole scientist, talks down to everyone, and sports a pointy beard and Russian-style fur hat to underscore a suspect patriotism. Carrington's scientific reputation based on his participation in building the A-bomb detonated at Bikini elicits no respect from the military men who remark, "a lot of good that did us." The questioning attitude toward nuclear (or atomic) power extends from the final scene of "Who Goes There?" in which the Thing's takeover of the earth is narrowly prevented as the Thing is about to make its escape, having built its own mini-atomic power generator. In 1938, atomic power seemed to offer its developer control over the world—thus the sense of panicked relief at the humble battery-generated incineration of the original 1938 Thing. After the US bombings of Hiroshima and Nagasaki and the subsequent development of Soviet nuclear ability, atomic power no longer seemed an advantage to the US military on the ground: the competition now emphasized containment, a strategy that relied on strategic geopolitical presence, not scientific-ballistic mastery.[26]

Yet, Carrington's desire to study rather than kill the Thing also represents the destabilizing potential of science. Only after the Frankensteinian monster has wrecked havoc—and attempted to kill the credulous scientist himself—does Carrington agree to help destroy it. In this version of the Thing narrative, there is no ontological uncertainty, no need for a blood test to establish who's who. The military and science merge to maintain threat—like Communism—as external to the state. Nevertheless, in a replay of the 1938 original's critique of the limits of science as a way of knowing, the Thing is destroyed as both threat and as object of study, even its saucer blown to bits by the incompetent military.

The major addition to the 1951 plot is "Scotty," the "dead weight" newspaper reporter, whose desire to break the story of the century makes his character appear foolish and even craven, perhaps as a reference to Byrd's faded glory and his former reliance on media and publicity. Yet, the reporter's status suddenly grows in the final scene. The film's conclusion shifts focus from the isolated and contained polar setting to address the paradox of maintaining both an informed citizenry and state and scientific secrets necessary to security. The key to this ideal of national security is not nuclear power, but good old-fashioned radio communication. The final scene, in which Scotty enjoins US citizens to "Keep Watching the Skies" has become a campy emblem of Cold War hysteria. In using film to retroactively assert the nation-building potential of radio communication, *The Thing from Another World* only amplifies the instabilities of national security within global expansion. Containment, a key Cold War strategy that called for maintaining crudely defended borders between communist and noncommunist zones, is thematized in this version of the Thing narrative. The polar setting suggests that a "freeze" in the destabilizing, globalizing tendencies of scientific knowledge and cultural exchange is a trade-off necessary for security.[27] The thawing of the Thing becomes then a symbol of the dangerous potential of internationalism. This theme is countered by the fact that Cold War geopolitics produced the context for the development of the ATS. The answer to the question: Why Be There? has been—since the Cold War—international science.

The "frozen laboratory" is one of the catchier descriptions of Antarctica, and indeed it does ring true to the ideal of objective, pure science. A lab is clean, controlled, ordered space. It is the opposite of the paranoid, spongy-bordered space of horror; it most certainly is cleaned of any "Russians...crawling around like flies." Hitched as it was to the wagon of science, human presence in Antarctica through the ATS developed seemingly outside the prevailing global forces. While the rest of the world underwent a Cold War, decolonization, and energy crisis, Antarctic politics ran on another track. International science stations proliferated and membership in the ATS expanded from 12 to include 44 nations. Environmental concerns also grew through the 1970s and became a major issue in the 1980s, as did the postcolonial challenge to the ATS by Malaysia to be allowed an equal share in future benefits of natural resource development, as well as the countervailing movement to name Antarctica a "World Park" stewarded by all nations and exempt from any development.[28] However, these challenges and proposals have not resulted in major changes to the integrity of the regime. Multiple factors account for the resilience of the ATS, and the regime certainly does have its critics. Overall, the idea of Antarctica as a continent for science has no serious political countermovement. Yet, scientific exceptionalism—the idea that scientific work is exempt from the political contexts within which it operates—like other forms of exceptionalism, in no way obviates the central problematic of national presence in Antarctica. Concern over human presence in Antarctica and its complex relation to geopolitics and to the political potential of scientific knowledge reemerges in the third version of the Thing narrative.

World Park? Or Office Park?

Thirty years after the triumphant (if anxious) containment of *The Thing from Another World*, John Carpenter's 1982 *The Thing* opens with a scene of the very skies that US audiences had been enjoined to watch by the 1951 film. A flying saucer crashes noisily into the white bottom of the globe as the words "The Thing" burn into the screen, a reference to the conclusion of the 1951 version in which the alien is killed by electrocution, a low wattage, distinctly nonnuclear (plot) device. Among the ironies of this self-conscious (re)opening scene is that the sky needed to have been watched 10,000,000 years ago, before humans existed. The impossibility of securing through science or government or individual heroism knowledge of human origins, or their future is immediately emphasized. Faithful to the 1938 original and borrowing features of the 1951 version, Carpenter stitches together a by now familiar tale of the discovery and thawing of the alien, its takeover of the isolated station, and its threat to the rest of the globe. The doom-laden conclusion, however, is Carpenter's own. If Byrd and international science under the ATS re-enchanted Antarctica as a space for a wary global peace-through-science, Carpenter disempowers both science and the military in a final scenario of mushroom cloud destruction of the Thing—and the station—as a pyrrhic victory. The final standoff between two insecurely human survivors is a nightmare of détente from which no future—and no remakes—can emerge.[29]

Carpenter's clever generic manipulation is to create a remake that despises imitation. Generation or reproduction is the problematic of the film (as remake), the plot (how to stop the alien takeover), and the setting (how to survive in Antarctica). Like the original story, Carpenter's *The Thing* mines the isolation and paranoia of a remote base, and like the 1951 remake, it references military characters and culture. But Carpenter's pastiche resists a clear historical placement.

Instead of Russians as threat, the inhabitants of the US base wonder "If we're at war with Norway" when a helicopter bearing Norwegian insignia comes out of nowhere to strafe their camp. The opening scenes emphasize Antarctica as a place of cultural and cognitive dissonance where the structures and expectations created for the rest of the globe do not apply: It is a place where nothing—or anything—can happen. But Carpenter's genre mash-up of horror and sci fi narrows Antarctic possibility into the surety of doom.

The coming apocalypse in this version of the Thing narrative is signaled from the opening scenes. The men on Base 32 (the numbered designation itself suggests order gone wrong, and a useless excess) are listless and unmotivated. Though each has a professional identity (cook, scientists, doctor, pilot, military boss, communications), they never once discuss why they are stationed in Antarctica or what their mission might be. As in a zero-sum game of chess, one by one the characters succumb to imitation or to the blowtorches wielded by their paranoid compatriots—to checkmate. Characters not torched must commit suicide to maintain the fiction of their singular, elemental being, an ironic device by which extinction secures identity. The theme of the war game resists both the rescue by science of the 1938 original and the anxious containment of the Cold War version. What we get instead is apocalypse born of a postscience, exhausted, "terminal," nuclear culture in the final scene in which the two survivors, sure of their coming deaths, nevertheless maintain a mutual vigilance as proof of their own humanity. They rescue humanity by insuring the isolation of Antarctica. *The Thing* makes use of Big Science's modes of communication, circulation of information and bodies, integration and assimilation against itself and for its own (inscrutable) ends. Rather than solve the problem posed by Antarctica's resistance to global capital, science and military along with nation and individualism are the very mechanisms of capital's self-destruction: Antarctica shuts down these possible futures.

Carpenter's is an antiglobalization, antiscience and anticapital vision (the two songs heard playing on the cook's boombox are Stevie Wonder's popular radio hit, "Superstition"—"When you believe in things you don't understand, you suffer" and Billie Holliday's "Don't Explain"—Carpenter's comment on the limits of rationalization). Killing off or discrediting all his protagonists, he declares an end to remakes, an end to progress, connection, circulation, and global knowledge. Aggressively retailing the hero-driven plot, Carpenter paradoxically decenters humanity in the very act of preserving it. Neither science nor good old American know-how prevails. Only a brutal containment can ensure the (suspect) originality of humanity. Antarctica is again a grand terminus—we are back to the eighteenth century, again. But in place of Cook's almost gleeful call for a rational end to territory, Carpenter's end is a dystopia of failed progress as the rational underpinnings of modernity undo their own foundations.

The sense of the impossibility of rational, linear progress comes through in the terminal setting of the Antarctic. Enclosure and paranoia come through in the depiction of the narrow hallways and crowded small spaces of the station, inspired by the military homo-social realness of the original story. Carpenter develops the paradoxes and ironies around authenticity and human culture in late capitalism through images of ennui and malaise. Characters pass the time watching reruns of TV game shows. "I already know how this one's gonna end," says one doomed character to his roommate, even though it is clear that he actually has no idea of what is coming, since he is caught literally in a (video) loop of repetition in place of horizon of the future.[30]

The sense of being nowhere and of going nowhere comes through most ironically as the viewer discovers that the entire action of the film—the discovery and thawing out of the Thing and its subsequent attack on its discoverers—has already occurred in the Norwegian camp. On a reconnaissance mission to the Norwegian camp, the US characters discover a scene of destruction, recovering taped footage of the Norwegian discovery of a Thing buried in ice. These scenes-within-a-scene echo the scenes of the discovery of the saucer in the 1951 version. Ultimately, what the viewer is presented with is a remake of a remake of a remake. When the origins are thus multiplied, only ends can be assured.

The motif of embedded quotations, screens, and scenes structure the self-knowing parody. In a key scene-within-a-scene, the base scientist Blair uses a computer simulation to describe the Thing's reproduction cycle and the likelihood of its takeover of earth's population in a matter of months. If the very modes of scientific method—deduction and modeling—can secure neither the primacy nor the survival of the human, the reliance on scientific reason may indeed undermine survival: Science itself may be the problem rather then the solution to the relation between humanity and Antarctic place. MacReady's blood test to distinguish men from Things—and his triumphant prediction, "Now I'll show you what I already know" (a near-direct quote from the 1938 original)—in reestablishing borders of species and individual, nevertheless suggests a devastating critique of science as a mode of knowing: He only needs to prove what he already knows, a narrative success that only undermines science's claim to disinterested new knowledge and objectivity. MacReady's test emphasizes the instrumentality of science to the state concerned with power, capital, and purely symbolic forms of colonization—all of the functions MacReady's presence in Antarctica exemplifies. As MacReady observes, "Maybe every part of it is a whole." He means to uncover a weakness in the structure and operation of the identity-less Thing. Yet, the opportunistic, every-part-for-itself Thing, rather than settling the difference between individual men and collective Things, activates the foundational contradictions of individuality, community, and nation as it reproduces itself through its own radical disintegration and duplication. But the Thing's method is really not so alien. Just as nation posits unified and naturalized origins, and national identity collates actually conflicting groups and individuals into a sameness it cannot in fact contain, the Thing reproduces its species as not-itself and the same simultaneously.

What does the triumph of the Thing's method of reproduction signify? Its modus vivendi is imitation—the refusal of difference. It produces copies. The Thing narrative implies the limit of the state as guarantor of order in Antarctica. Every part as also a whole Thing in itself, refusing both internal and external difference, also describes the foundational ethos of an international agreement among supposedly equal states, such as the ATS. Since the ATS is an agreement among states to curtail and significantly contain and defer their rights and powers, it is especially vulnerable to non-state or trans-state actions and phenomenon; in other words to Thing-ing.

The Thing is not a thing proper; it is not an entity but rather, a process that uses imitation as a form of production. Exploiting circulation, mobility, privacy, belief in originality or authenticity, those very means its human hosts used to establish a proto-civilization in Antarctica, the Thing calls out the constituting contradiction of nation as an imaginary unity of individuals who would otherwise notice that their histories and interests are not in fact the same. As the unity of the group breaks down, Carpenter helps us understand a US strategy of neoimperialism, of power without territory, that makes post-WWII US policy predictive of its new hegemony in Antarctica.[31] Neoimperialism is an understudied feature of contemporary Antarctica that bears structural similarities to neoimperialism elsewhere on the globe today. These flexible, mobile, and extremely adaptive (imitative, Thing-like) entities position themselves to take advantage of local conditions, whether it be war, cheap labor, untapped resources, or more open banking laws. These corporations destabilize local economies to profit their remote headquarters, and close down production when conditions are no longer favorable.

One such corporation, Raytheon Polar, a division of Raytheon Corporation, a well-established US-headquartered international weapons builder, has been since 2000 the major contractor for all US Antarctic support services. Based in Denver, Raytheon Polar sub-contracts almost all of the labor in support of the National Science Foundation's (NSF) Polar Programs. Although the Treaty required the phasing out of purely military occupation and goals in Antarctica, its military history continues to shape life on the continent in the form of infrastructure, personnel, and transcorporate relations and policies. Demilitarization has not resulted in an absence of military in the Antarctic. Rather, the aftereffects of militarization can be traced in Antarctica's built environment, human culture and language, as well as its infrastructure. Science too is inscribed with military traces and histories: It can never be completely separated from its military modes and contexts.[32] Raytheon's role as a subcontractor raises these complex issues. Raytheon's Antarctic wing is a small part of its global business and is far from profitable. Rather, the company promotes its activity in Antarctica in terms of prestige, not profit.[33] This lack of actual profit lends itself to a reading of Raytheon as part of a neoimperial project in which profit is not directly necessary to under-score presence of capital. Nor is Raytheon's role directly military: Its Antarctic activities constitute a new form of US empire in Antarctica. While on the one hand Raytheon's Antarctic presence can be seen as merely performative of a patriotic service to science or a quaint notion of exploration history, another way to understand US presence in Antarctica is suggested by Chalmers Johnson's observations that "vast network of American bases on every continent except Antarctica actually constitutes a new form of empire—an empire of bases with its own geography not likely to be taught in any high school geography class."[34] That Johnson cannot see McMurdo Station as part of this vast network says more about his training than the more obvious fact that McMurdo Station too fulfills a neoimperial, post-territorial stra-tegic position for US global hegemony.[35]

What does this say about the role of the ATS? The ATS allows an array of troubling activities under the aegis of science.[36] The ATS works by not protecting all possibilities and futures in Antarctica, but only one mapped (historico-spatialized) projection of Antarctica, one that most benefits neoimperial powers, particularly the United States. What kind of future does this analysis of neoimperial processes and practices in Antarctica under the national science program–dominated

ATS predict? Well, not necessarily a dystopian one unconnected with that of the rest of the globe. The nature of the shifting ATS suggests that it can be a working descriptor of change that will protect Antarctic nature. If the changes of the 1980s brought on by nongovernmental organizations (NGOs) such as Greenpeace and non-claimant and non-acceding states to the ATS can be a model, the ATS may be able to accommodate alternate visions of Antarctica's relation to the globe, such as the World Park concept. Is this a likely scenario, given present indicators and scientific culture? Science itself is a multiple enterprise, as the so-called debate over how to apply data on climate and temperature shift demonstrates. For the time being, the unimpeded collection of that data will be a more concerting issue than future resource extraction.

Byrd's messy origins of colonial, personal, and global vision have never been resolved and we see their inconclusive developments in contemporary Antarctica as it is internationalized and instrumentalized through a complex arrangement of states and non-state actors, especially the transnational corporation Raytheon. The possibilities for Antarctica's future under the ATS develop only through the deferral of certain futures (of territorial possession and resource development) in favor of a present of presence through the modes of science. Raytheon provides for the contemporary period a quasi-military infrastructure to replace the scientific self-policing of the 1938 original, the citizen watch of the Cold War version, or the depoliticized, routine-numbed nihilism of Carpenter's Base 32. Raytheon, the United States, and NSF work together to create a postinternationalist, post-postcolonial future: Antarctica as a workplace. Where Byrd's singular heroic encampment failed with his body, and military presence proved too volatile for international agreement, the joining of national science and transnational corporations has succeeded. In this mapping, capital and nations reach a new standoff, or another deferral of a future, which the ATS has come as much to prevent as to produce.[37]

Perhaps, it is best to conclude by reengaging the paradox of generation at the core of the Thing narrative: Through imitating, or remaking, it initiates catastrophic change. The "neo" prefixed to imperial similarly indicates both imitation and difference. It is imperative that scholarship approaches the partial, particular reinvigoration of empire occurring today in Antarctica. Australia, Argentina, France, Norway, Britain, and Chile maintain claims that predate the ATS. Britain has reasserted its claim to the United Nations, in seeming violation of the ATS, and angering and alarming numerous signatories. Russia has advanced its plans to drill deep into the lake that gives its permanent base Vostok its name, despite concern from the international scientific community. And the United States has rebuilt and extended its Antarctic empire—without territory or sovereignty, or economic return on investment. Today, it remains impossible to answer the question science fiction asked—Who Goes There?—with any clarity, much less finality. Nation, science, and empire shift methods and strategy, and reconstitute themselves as continuously as does the ice itself. And the ice itself—the living ice and the living on ice—continues to change bodies and minds even as it responds to the control and management of humanity. As the numerous new claims and the explosion of very different types of science (or quasi-science) activity in Antarctica attests, the ATS will still operate even as the ice shifts because its foundations are as shifty as the space it describes. Structured as numbered amendments, the ATS can change to accommodate novel technologies and circumstances. And yet, its principles, because they are so broad and simple, a series of can and can't-dos, will continue to describe the horizon of human possibility in the poles, developed as they were from a history of just the sort of human activity it describes. The ATS is a spectacular limit, a self-fulfilling prophecy of just what can and will constitute failure and progress in Antarctica. It allows an array of troubling activities under the aegis of science. It is a limit without limits, tied as it is to science, the imperial discourse of the future and the possible—in space and in institutions.

The final section furthers the analysis of the effects of big science—meaning international, interdisciplinary, and logistically integrated science of the kind currently operating under the ATS—on the "living ice" of polar work culture.

Living (on) Ice

One counter to science organization of Antarctica comes from within its very management system: the subcontracted science and support workers. Their experiences, cultural productions, including built environment and visual art, books, and blogs, as well as their legal and everyday acts on and off ice constitute a counter archive within the veritable belly of the beast that is NSF-run Antarctica. US Antarctic workers may never become citizens of the ice, and are in fact prevented from extended and uninterrupted work contracts by the imposition of a seasonal labor cycle. Yet (along with the worker population of other ATS bases), their substantial and growing subculture is probably the most significant, exponentially growing and understudied living archive of Antarctica.

Antarctic work conditions receive little scrutiny for some very distinct reasons of Antarctica's little-known status and extreme underpopulation. US workers tend to be recruited in a tight word of mouth circuit in cities where subcontractors like Raytheon are based. Classed values of allegiance to regions and the corporations that have helped build them, white working- and middle-class cultural isolationism and privilege, and class elitism and exclusion borne of special relations to the military, education, or other entitlements, not to mention that just plain restlessness of middle-class white youth and professionals drawn by the advertisement of working in the most unusual and "exceptional" place on earth has built a steady flow of applicants willing in many cases to work menial jobs just to be able to be in Antarctica.

In an era of environmental consciousness, the promise of a pure relation to nature is a primary appeal as well. And certainly, many come to escape their routines and the dullness and dead ends of even professional labor. They want to be special. In a way they are, but not in the way they might have fantasized before signing up. Working on an Antarctic base is a combination of military duty, community college, and an expat community living relatively well in a country made unstable by forces in part attributable to the dominant nation of their origin. Those intensifications of emotion and expectation around working in a natural wonder or a place of prolonged cultural fantasy must be tempered by what are some of the more mundane aspects of working in Antarctica: Ultimately, cleaning toilets in Antarctica is cleaning toilets. It makes more sense to understand workers at the US base as offshore office workers or as migrants, following a job site. Or, they may be in a strange twist more similar to the laborers used by a transnational corporation coming to exploit cheap labor. The biopolitical management of US workers in Antarctica, far from being exceptional, is extensive with the flows of migrant labor, cheap labor, and what is now termed affective labor, in which the negotiations of the worker, the corporation, and state feature forms of investment of emotion and belief that contribute to structuring profits away from labor. People are willing to be exploited and controlled due to their investment in the aura of distinction they perceive in being one of the infinitesimal people on earth to actually live in Antarctica.

The fact that so few people have or perhaps will ever experience Antarctica directly intersects with Antarctica's visual culture in a counterintuitive way. Rather than being the least represented place on earth as a function of its few people, Antarctica is arguably the most overrepresented place on earth, with more images produced in relation to population than anywhere else on earth. Fewest eyes have seen the most Antarctica—and reproduced and circulated what they have seen. What this strange situation adds up to is my claim that Antarctica is the most mediated place on earth. It is the most mediated in terms of the amount of mediation necessary to circulate on-site capture to the bulk of humanity who will never visit the continent.

Antarctica brings out a consuming frenzy to document—human time there is so fleeting and oversold. I met many enthusiasts in my six weeks during the 2004–5 season at McMurdo and various field locations. Cameras were everywhere; the Internet and the point and shoot had transformed being in Antarctica more irrevocably than the much more lamented arrival of women into the boy's club that began in the 1970s in earnest (and before that science, which ended the rule of the military and cued a homosocial war between soldiers and scientists central to the Cold War myth).

McMurdo hosts its own film festival its own workers produce in their spare time. McMurdo also has a public H-drive on which anyone can post their images of the Erebus volcano—or of the drunken debauchery of the previous evening. Policing the H-drive became a major concern of Raytheon's Denver headquarters, a concern not atypical to any corporate enterprise's public relations and human resources divisions. My interest is in the stakes and effects of the use of the H-drive for the workers. Were the McMurdo workers engaging in the kind of auto-ethnography prevalent in parallel cultures of schools or workplaces? Or in volunteering to post their exploits on the public H-drive, did they anticipate the viral explosion of social networking of Facebook and what is now referred to as life-logging, or were they feeding into a larger system of surveillance and legal management? Are the McMurdo workers expressing their freedom or are they enacting or advancing their very culture of control and management?

This question of the creation and circulation of affective flows of belonging to Antarctica through labor comes to a head in the McMurdo Station workers' 2005 Gulf War peace protest. Although far from the mainland United States, Raytheon employees, like military personnel stationed throughout the world, have citizenship rights, including the right to vote. Raytheon has an investment in curtailing and controlling compensation and work conditions. A court case was finally decided against Raytheon employees who argued they were not working in US territory and therefore should not be subject to US taxation. Curtailed and controlled citizenship is the hallmark of labor in the US Antarctic; the rights of workers to express themselves politically as well as to control their access to and production of information have been an area of ongoing contention. In the political climate of the Second Gulf War, McMurdo workers began circulating in January 2005 a photo of workers forming peace symbol with their bodies set off against the "blank" ice. The Antarctic peace symbol joined other similar images of people in multiple locations protesting the war. What seemed new was the unusual setting—the forgotten Antarctic.[38] As a protest designed and executed by US workers within the highly controlled environment of Raytheon subcontracting, however, the protest and the image they produced and circulated is much more than a visual-media integration of Antarctica into global circuits and politics. It is also a very carefully designed image shot from a distance to protect individuals from reprisals and to cynically use the bulky clothing issued for safety as a cloak of anonymity for protesters.

Combining bodily process and manufacture, anthropomorphic attachment to and integration with ice, and nostalgic throwbacks to US 1960s' Hippy culture, the peace symbol circulates and encircles even as it surpasses the individual, the human, the orthographic in returning or reproducing Antarctica to its earliest fantasy as cartographic blank or white page. Widely circulating on the Internet and included in the exhaustive list of global protests 2002–3 archived on *Wikipedia* (http://en.wikipedia.org/wiki/Protests_against_the_Iraq_War), the McMurdo peace sign is a pixilated, animated, circulating, and more convivial vision of life on and in ice ironically composed of supine, dead-looking human bodies. In a further twist, it is a peace protest from the land that supposedly has never known war, bringing it full circle to Byrd's rescue of his failed nationalism through a utopian vision of international peace.[39]

It is this conventional break of "post" war that often characterizes the Cold War era in which the IGY and ATS codeveloped that no longer holds. The rationales underpinning clear borders of periodization and nation, as well as those between militarization and everyday life have dissolved. Global war sustained through capital flows now describes the condition of globalism itself; there is no clear border between war and peace, just as there is no border between state military and private corporate power. If conventional geopolitics argues that war brought "peace" to Antarctica after WWII in the form of the IGY and ATS, today that peace is war in Antarctica.

Chapter 4

On the Road with Chrysler: Virtual Capitalism and Empire without Territory

Antarctica has posed a problem for capital in two contradictory ways. One, in that it is difficult to profit from barren ice; and two, in that it is even more difficult to account for the many ways that capital has in fact profited from the idea and the material facts of Antarctic territory, including the very idea of its unprofitability. The first, few people would argue with as there are countless laments to the fruitlessness of ice and to its limited possibilities for capital and empire. The second is a more interesting problem, suggesting that valuing Antarctica is as much indexed to the representational as well as logistical systems that frame Antarctica—such as exploration and empire or cartography—as it is to the Antarcticas that are so framed. Extending the previous chapter's discussion of US official exploration and policy from the 1920s through the era of the ATS and its present science–corporate management, this chapter looks more closely at the virtual and actual infrastructures of capital involvement in Antarctica. While contemporary Antarctic development may appear to be broadly humanist, benignly scientific—and with the establishment of a "road to pole" even quaint and fordist in its production mode—the government–corporate advertising strategies and built environments in Antarctica constitute a new form of empire and possession of territory without territory.

Chapter 3 argued for the way that Antarctica, far from being a territory of heroic past or of a technoscientific future, has been part of global flows of unequal power. Taking the joint US–Norwegian "Road to Pole,"[1] a 995-mile traverse of compacted snow and staked flags between US McMurdo Station and the US Scott–Amundsen

South Pole Station that became operational in 2010, as both symbol of connection and as a material feature of human presence in Antarctica, this chapter develops how corporations strategize advertising campaigns to profit from Antarctica despite the Treaty's strictures on territorial claims, military presence, or economic development. It departs from Stephen Pyne's provocative formulation of Antarctica as an "information sink" (*The Ice* 20) where expeditions, men, attention, and money all flow into Antarctica to be absorbed by its various lacks—of history, people, resources, to consider how science support infrastructure produces a new form of value and territory.

A Great Location for an E-Marketplace

2000 was a time of unwarranted exhuberance for the powers of the Internet to create new spaces for capital. IBM's full-page advertisement appearing in the print version of the newspaper was certainly striking. Its headline, "What a great Location for an E-marketplace," seemed to provide a horizon for the light blue expanse of the scene. A partially depicted prow of an ocean-liner-type ship on the right side and the tiny form of a lone penguin at the bottom help to create the impression of territory in this otherwise vast and empty expanse of ice. Small print and company logos lining the bottom of the layout await those sufficiently intrigued to want to know the details of IBM's doomed business plan.[2]

In retrospect, after the bursting of the tech bubble, IBM's vision of Antarctica as "a great" location for an "E-marketplace,"—a space of what Karl Marx called "primitive accumulation," or as the permanent outside of the capital system—seems quaint. Like Berlin's Fernstrum or Buckminster Fuller's geodesic domes, IBM's virtual marketplace in Antarctica offers a melancholic futurity—an "obsolete future"[3]—haunted and controlled by the past.[4] While IBM's marketplace on ice failed to catch on, the company's image of the lonely penguin juxtaposed with the icebreaker representing civilization links back to countless photographs of the arrival of the ships of the Heroic Age to the ice. More importantly, it anticipates the 2005 animated film *Madagascar*, in which penguins exiled in a NYC zoo seek to return to their "homeland" in Antarctica. A much-circulated scene of the exile's return echoes the IBM advertisement's layout of vast, horizonless ice, prow of the icebreaker and a lone penguin, who after a few long moments in which is heard only the howling wind of this clearly uninhabitable location, declares "This sucks." Leaving aside the parody of nativism's idealization of the return, and even a bit of shaudenfreude around IBM and other corporations' miscalculations around virtual marketplaces, the unstated irony of the depiction of virtual means to circumvent Antarctica's geophysical resistence, is the ongoing investment in building actual, physical roads (if not marketplaces) in and to Antarctica.

Many worry that projects like the road to pole are incompatible with the stated ideals of the ATS to protect Antarctica from undue development and from environmental distress.[5] The ATS defers states' territorial claims into the future in exchange for relative freedom to act and develop under scientific management. Sanjay Chaturvedi argues that the Treaty extends a nineteenth-century model of space through a "legal freezing [of territorial claims that]...protects and promotes a particular vision of the continent anchored in the colonial past" (134). While I share some of Chaturvedi's and Klaus Dodds's (2006) concerns, the claim that empire is being reasserted loses its accuracy in replicating an older postcolonial model of thinking based on conventional notions of territorial claim and occupation that cannot account for the specificities of ice and that misapprehend the ways that settler colonialist repertoires and technologies have escaped the discursive tenacities of legal and historial regimes.[6]

Territorial empire is not being reasserted in Antarctica. Rather, it is presently occurring in a neoimperial occupation and appropriation of Antarctic space, if not territory. If anything, rather than hang on to a frozen past, corporate image strategies bypass the past, even if the language of time and timed event persists in talk of "firsts" and of a linear aiming for a future. By not understanding the proliferation of stations and permanent populations across Antarctica as a species of a new land grab those who purport to speak for the continent and the highest goals of enlightenment science and progress and international fairness are missing the bus, or should I say, the road that has replaced the territory.

Road to Pole

Presence—corporate or corporeal—in its traditional form has relied on the bodies of explorers and the workers who carry on and replace them on the continent. However, under the present regime of non-claimancy, the indexical forms of empire discussed in Chapters 2 and 3, which are reliant on the body of the explorer and the trace on the ice, have been entrained into a new type of indexicality that uses movement and process rather than fixed or permanent location. From the respatializing of the continent produced under the ATS emerges the reterritorializing of the new road to pole. The "road to pole" supply chain laboriously marked out and maintained by a transnational team solves the problem long posed by article IV of the Treaty deferring claims: the road to pole—its management and constant upkeep as well as its imagined trajectory of movement—replaces the territory itself. This new territorialization without territory can be understood only in part as motivated by rising fuel costs and the anxieties of the end of oil capitalism. More significant is the maintenance of the corporate–logistical road as it retains the affective trace of the masculine heroic national laboring body while exploiting economic crisis to become a tool of new forms of possession of territory (and dispossession of laborers' rights) whose supply chain also includes the production of an idea of the future itself.

In 1963, Volkswagen (VW) proclaimed itself the first car in Antarctica. The full-page black and white photograph is dominated by a head-on view of a dark-colored VW bug amidst a sloppily barren ice scape; shot from beneath the level of the vehicle, the bug seems ready to roll, though no actual road is evident. Wearing polar gear and yet leaning out of the open window to wave, the driver appears discordantly jaunty yet also distant, in another world, a bit like an astronaut. The tag line beneath reads: "The first car at the bottom of the world."

While the VW may have been the first passenger car manufactured for the paved roads of the developed world and brought to Antarctica, it was far from the first vehicle. Mechanized transport had been leaving tracks on ice since Scott's tractors fell through the sea ice in 1907. But the German corporate-Australian national product was the first contract car in Antarctica, shipped to Australia's Mawson Station as a boxed kit with spares in anticipation of the inevitable failures. The license plate reading "Antarctica 1" inserted the VW bug familiar to middle-class viewers all over the world to the new terrain of Antarctica. That there was no corresponding national road "1" in the nonnationalized territory only strengthened the car maker's claim for its versatile and hardy bug. VW's traction on ice just a few years after the ATS came into effect in 1961 is an early instance of the pattern that now characterizes both the scientific maintenance of Antarctica's built environment and its management. The self-promotion of the advertisement loops into the anticipatory maintenance of its own risk. This proto-disaster capitalism, ready to profit from even its own failures, characterizes the complexity of virtual and material scales of Antarctica's management under capital.

In 1998, Matthew Coolidge, an artist and founder of The Center for Land Use Interpretation (CLUI) created an image also titled

"Antarctica 1" quite distinct from the product-oriented advertisement image. It was a "found" image of an overview of McMurdo Station showing the network of industrial–military service roads and numerous vehicles—none of them VW bugs. CLUI's "Antarctica 1" also indicates a road that in fact is not (yet) there; it depicts the base or destination of service roads, but not a road in itself, certainly not a traditional paved road. A road in Antarctica does not even resemble an "off" road, familiar to biking enthusiasts. A road in Antarctica is in reality a set of staked flags, harder to keep groomed than all the green lawns in the water-parched US Southwest. In the 35 years between VW's "Antarctica 1" and CLUI's 1998 overview of McMurdo Station, the road has become, if not a long and winding reality, a specified starting point. This road points back to a past of linear, arduous trekking between the continent's shore and the pole, as much as it indicates a future horizon.

Virtual Capitalism

The footprints of the heroes were indexical marks of firsts and national accomplishment. But as earlier chapters discuss, they were not necessarily connected; they did not add up to roads. The roads required by present-day national science installations are built by workers, who are hired by subcontractors. A series of 2009 DC Metro ads "for a product no consumer can buy" by military subcontractors vying for the symbolically prestigious if not lucrative Antarctic support contract with US-NSF give a sense for the way heroic imagery is retread to support the logistical demands of neoliberal transcorporate capitalism. The roads to pole built and maintained by national science programs are roads no commuter could travel. They depict bright skies and clean white bergs floating in a blue sea. People wearing bright red technical gear gesture from their rubber zodiac in professional ways to the horizon. The images suggest Boy Scout adventure more than military subcontractor science support as they connect Antarctica's idealized wilderness to more traditional or familiar zones in the US West of the Rockies.

Corporations advertise their ability to develop and maintain business and national (science) projects in an extreme environment, needing military-like logics and capability. The advertisements' potent mix of nationalism and environmental universalism appeals to Beltway workers (not to mention politicians voting on the contracts) riding the Metro. Military contractor Bill Bodie of KBR, Inc. distinguishes between his corporation's pursuit of proposals for support of

"U.S. forces in Kuwait, Turkey, and Spain" and the National Science Foundation support proposal which lies, he says, "[o]utside the defense arena." Bodie then describes the NSF support contract as "LogCap" [a logistics augmentation contract that KBR won for the US Army] "on ice," a comparison that belies his own distinction between military and nonmilitary environments. He goes on, confusing the distinction he attempts to impose between military and nonmilitary support, observing "Antarctica is not a hostile environment, at least in terms of people shooting at you, but it is very remote, very austere, and the actual work requirement is to ensure that scientists down there are taken care of just as soldiers are taken care of" (*ExecutiveBiz*).

In her 1987 essay "Heroes," Ursula Le Guin defended a then very unpopular Sir Robert Scott, who, as I discuss in Chapter 2, had become the type of the imperial oppressor, derided for his British Navy ideals as much as for the logistical failures and errors of his expeditions. But Le Guin defended his approach to Antarctica as a place, especially in his avoidance of casting ice as the enemy, an argument that long preceded the very recent recuperation of Scott as a scientific explorer more in step with today's international science regime. Corporations, however they try to project an environmentalist–adventurist image for Antarctic endeavor, revive the militarization of ice and the casting of ice as a hostile enemy. That much of present-day science in Antarctica is driven by research on climate adds layers of irony to Bodie's connection of KBR, Inc's "shooting" work to its logistical support in Antarctica. The advertisements use beauty and humanism to paradoxically anchor a non-sovereign deterritorialized relation to place that hollows out the very histories of empire and industry its image repertoire invokes. The corporate use of Antarctica alongside its more directly profitable, or in the case of KBR, Inc and Raytheon Corp., directly military divisions, is a crucial arena for the renegotiation of power, biopolitical management, and the reterritorialization of traditionally US space and Antarctic ice. The next section looks at an earlier advertisement campaign featuring Antarctica by a multinational corporation whose deep cultural connection to US nation through a domestic car culture makes for an even more striking demonstration of how corporations profit from Antarctica not as territorial possession, but as unadorned ice.

On the Road with Chrysler

In a well-known 1950s television advertisement, Dinah Shore invited American viewers to "See the USA in your Chevrolet," thereby

linking the post-WWII interstate highway system, fordist automobile production, and nationalist discourses to a specifically American
landscape. Aligning individual, nation, and land as integral features of
a naturalized national consumer order, this Chevrolet advertisement
demonstrated the marketing of the American Dream in the postwar
era. Such a strategy continued to dominate automobile advertising
throughout the 1970s and 1980s, even as US automakers began moving their production sites overseas to take advantage of lower labor
costs and to avoid expensive negotiations with powerful unions. Not
until the 1990s, however, did nonnational image repertoires begin
to appear in automobile industry ad campaigns. A much-circulated
1994 advertisement for Chrysler Corporation depicts the continent
of Antarctuca from the vantage point of space, zooming through
the ozone hole directly onto the South Pole, from which a cartoon
penguin gazes up plaintively as if to say, rescue me. The purpose of
Chrysler's advertisement was not directly to promote its products,
but rather its adherence to standards of pollution control, stating:
"The Ozone hole has protected us for 1.5 billion years. It's time we
returned the favor."[7]

Chrysler's use of Antarctica is a telling instance of the way nationalist imaginaries are continually being refigured by and through the
rise of transnational corporations (TNCs), whose modes of operation
are not once and for all settled in particular nations or locations, but
rather, like weather and climate conditions, constellate and move across
borders. Neither permanent nor ephemeral, neither wholly unattached
to national formation nor entirely subject to nation, transnational corporate modes of production both respond to and shape economic and
cultural resources "in situ," so to speak, as opposed to in fixed place.
No longer does "buy American" strike consumers as a necessary and
radical act for claiming US world power by consolidating US workers and consumption; in fact, the slogan has been rendered materially
incoherent by automobile production practices such as subcontracting for parts and foreign factories that obscure the relation between
national labor and national brand.

Much of the newfound comfort with the transnational can be
attributed to its everyday presence in information circuits and recent
ad campaigns which link televisual, print, and electronic modes of
representation. Yet, as a technology in practical use by corporations,
governments, and individuals all over the globe, the information
highway does indeed extend beyond political rhetoric. As the "information highway" metaphorically and materially merges the old US
interstate highway system of the postwar era with the global cyber

sphere, US TNCs such as Chrysler boldly market the international nature of production. Such marketing representational strategies seek to resolve tensions between the nation and the TNC within the larger context of the globalization of commodity culture.

In its focus on the nonnational and exotic Antarctica, the Chrysler advertisement stages its investment in a postnational imaginary by highlighting the positive environmentalism of postindustrial production. By reading the Chrysler advertisement as evidence of the representational shift toward global consciousness as a marketable strategy as early as the mid-1990s, this chapter has considered how the unprecedented global arrangements, both material and symbolic, of the TNC have produced a "virtual empire" of value and meaning unavailable and perhaps unthinkable under the rule of nation. It is this virtual empire that replaces the colony as the material location for imperial accumulation, posting the nation and insuring the ultimate survival of the TNC to rescue the globe, as the advertisement promises, "from a problem hanging over all our heads."

Rescuing the End of the Earth

One of the most useful strategies of the TNCs in their search for ever new markets and to integrate and reorient existing ones is to marshal consumers under a sufficiently global purpose. To this end, the problem of global environmental degradation serves well: A global problem demands a global actor, and the TNC is well invested in taking up the global position vacated by the fall of national superpowers. But how can the TNC dare be so cheerful and optimistic about what looks like the end of the earth? In part, it does so because that ending enables a revamped consumer protectionism that shifts the power of oversight and perpetuity from the state to the TNC itself. It is thus no surprise that Chrysler has been especially drawn to the environmental trope. In fact, the Antarctic advertisement is part of a larger environmentally focused Chrysler campaign that ran in 1994–5 nationally in magazines such as *National Geographic*, *Time*, and *Newsweek*. The ads all share a similar format with a large color image of a natural phenomenon or of animals and accompanying public-service style text touting one of Chrysler's environmentally friendly policies; none depict humans, buildings, or machines. In one advertisement, arm-like extremities (flippers, tails, trunks) of exotic animals from across the ecosystems join in a circle as if in preparation for a team cheer. In the good fight for the preservation of the natural world, Chrysler acts here as the inaccessible agency offering a helping "hand" to endangered species. Chrysler's projection of its internationalizing prowess is imaged through the seemingly universal aegis of the natural world, whose anthropomorphized concerns most clearly cross national borders by inviting global claims.

For my purposes, of course, the advertisement that focuses on the globe from space offers the most striking enactment of Chrysler's transnational pursuit. As I suggested at the outset, there is something deeply significant about Chrysler's appeal to global consciousness via an image of the environmental degradation of Antarctica. With no indigenous population or habitable landmass, Antarctica defies the reigning logic of nation: its territory is not yet divisible into nations and it has no human culture to register it as a place. Even the cute penguins and seal of the image live on the peninsula, well off the center of the continent. Contrary to the advertisement's jaunty picture, the bulk of the continent is a frozen desert lacking access to free-flowing water.

As symbolic and material image of the "end of the earth," Antarctica certainly works as a focus of environmental drama. Each week, it seems, *The New York Times* reports the latest oil spill, mass bio death, and widening or closing of the ozone hole. Every footprint marring the surface of the ice—from those left by dinosaurs roaming a once-tropical southern landmass to explorers and now the latest tourists—testifies to the incompatibility of human and Antarctic time and place. Chrysler's advertisement for the continent plays off of this incompatibility by, on one hand, inserting Antarctica into public consciousness as potential resource and, on the other hand, promoting the power of TNC capital to incorporate the marginal, to recuperate the wasted, and to heal the damage begun by formerly national-based industrialism. Chrysler's map of the south recenters the globe around Antarctica, populates (with animals) the uninhabitable, and projects conventional life onto the radically other. In short, the advertisement successfully absorbs Antarctica into the global economy as preserve and as territory—and of course as material place in need of the kind of capital and reform that only the TNC can provide.

In troping the posthuman exploration image of the globe seen from the comforting distance of deep space, the advertisement gives readers a history-less globalism: "The Ozone layer has protected us for 1.5 billion years. It's time we returned the favor." Just as Benedict Anderson writes in *Imagined Communities* that nations must justify themselves as ancient, Chrysler too imagines this newest continental territory as ancient—1.5 billion years ancient. In the same way that the narrative of the discovery of America by Americus Vespucci or

Christopher Columbus, or even by Pacific Islanders reinforces the naturalness of America as a place and a nation, we find that Antarctica has always been there, waiting for Chrysler to represent it. By narrativizing the history of place that has led to this moment of mediated representation, Chrysler creates Antarctica as the newest global entity with a recognizable history which, though suppressed, seems only to require for its fulfillment the simple reciprocity of a business deal.

But this transnational representational regime, smooth as its appeal may be, does not entirely avoid contradiction. A TNC symbolic is just as fictional as a national symbolic. Even within the TNC network of the advertisement, nation intervenes: Chrysler, the ozone-destroying avatar of fordist production, comes to the rescue of its own creation of Antarctica as an otherworldly, nonnationalized and nonindustrialized place through its zealous compliance with "government guidelines" on the atmospheric pollutants, chloroflurocarbons. Nonaligned scientists consider the government guidelines insufficient in the first place. Not only is the logic of Chrysler's appeal faulty, but the rescue too is a sham, as the TNC rescues us from a problem it has produced, thereby securing the inevitability and necessity of its interventions. In addition, the simple business deal proposed by the advertisement, a deal that enlists the reader's help in the rescue of the continent has an even more contradictory reference: the 1982 Chrysler "bailout." At one point in its transition from national industry to transnational and diversified entity, Chrysler, the sick corporation, needed the help of nation in the form of government-guaranteed loans; presumably, through the promise of association with Chrysler's powerful transnationality—that paradoxically figures as a national reinvigoration—the nation has been repaid.[8]

Chrysler thus produces both problem and cure, along with the very place on which it stages its rescue drama. Chrysler's vision of Antarctica as the last place on earth is also the first place for the technological development of new representational practices and new ways of tying the unusable (or wasted) earth to the engines of capital accumulation that have propelled the world's economies into transnational exchange. It is only the combined virtual and material operations of the TNC that have delivered Antarctica from margin to center and into profitability.

The United States is committed to the ATS, seeing it as the best policy for insuring future possible territorial rights and scientific access in the present. Policy makers also see the ATS as the best preventative against the possibility of nationalist military conflict (Joyner and

Theis). Chrysler, thus, has pulled off the capitalist impossible dream by creating value from a wasteland in a global economy marked by increasing stricture and scarcity. Moreover, it circumvents the problem of ecological degradation that it congratulates itself for solving by extracting value from Antarctica without leaving a single step on the continent itself.

The cartoon fantasy of profit without destruction or of a self-regulating and cleansing corporation has a strange correlate in landscape photography aimed at conservation. Tending to remove or otherwise demote and control the mark of civilization, even to the point of obscuring the effect of the photographer, the tradition of landscape photography can play into the suppression of the political contestation over the land depicted by creating an illusion of autonomous nature. Chapter 5 extends a concern for corporate stategies of occupying territory without sovereignty in the work of four major visual artists, Herbert Ponting, Eliot Porter, An-My Lê, and Robert Smithson.

Chapter 5

Photography on Ice

> *As world reserves of oil and gas go on shrinking, and as the richest mineral deposits approach exhaustion, international consortia will begin to exert pressure on governments to permit exploratory drilling in the unglaciated dry valleys...and on the continental shelf of Antarctica.*
>
> Eliot Porter, *Antarctica* 1978

Given the ongoing economic and environmental upheaval, US conservation activist and landscape photographer Eliot Porter's concern 30 years ago over the possibility of pressure from a "consortium of petroleum corporations" to drill in the fragile US Arctic might well have been written in the summer of 2011, as the ice of the North Pole melted to create a new, open zone of imperial contestation. Or it might be written in the near future, to decry new zones of economic development that restructure the Antarctic for exploitation. Although protected from economic mining interests by the 1959 Antarctic Treaty for the present, Antarctica has nevertheless become a focus of economic pressure driven by a renewed scramble for scarce resources.[1]

This chapter considers the linked shifts in US conservationism and landscape traditions as they have themselves contributed to the remapping of Antarctica within global networks of economic and military investment. In tracing a history of landscape photography in Antarctica from Herbert Ponting's Heroic Era through US government-sponsored photographers Eliot Porter and An-My Lê, this chapter demonstrates the way Antarctic ice has been blanked out and filled in and blanked out again, leaving it available in a deterritorialized future instrumentalized by neoliberal militarism and permanent war. The conclusion reevaluates the commodifications of ice in and outside of the culture industry through a reading of Robert Smithson's photostat collage "Proposal for a Monument in Antarctica" (1969).

In pointed contrast to Porter's written concern over oil capitalism's potential to make a ruin of ice, his photographs comprising *Antarctica* depict a "timeless" and composed Antarctic landscape, betraying no sign of human intervention. Porter, a celebrated nature photographer, was the first to bring to US public attention a place very little known. Much wider and deep public cultures of Antarctica developed in Britain, where the race to the South Pole at the turn of the twentieth century reflected directly on imperial history, and in southern hemispheric colonial and commonwealth countries like New Zealand and Australia, where relation to the British crown still shapes national self-understanding and the southern hemispheric-ness intensifies a pattern of cultural belonging absent in Europe and the United States. "Deception Island" serves as the cover for *Antarctica*, emphasizing the relative familiarity of its greenish, grass-like lichen and varied rock forms, and clearly echoing the strategies of Porter's earlier landscapes in the American west. Porter helped create the Sierra Club aesthetic that was also a complex strategy of promoting conservation in the US postwar era.[2] Advocates sought to spare Yellowstone Park development. But advocacy required promotion. The photographic school that developed to capture and control the image of the Western landscape created dramatic and anachronistic images of seemingly untouched nature. The tourism unleashed as a force to save the park also created a pressure to industrialize and pave roads through the newly desired wilderness: a paradox of ecotourism now continued in Antarctic discourses. Following the lead of the Sierra Club to aestheticize as a strategy to promote conservation, Porter aimed to preserve the Antarctic wilderness against the inevitability of capital through the power of his lush, full color coffee-table book. But bringing the little known, nonnationalized Antarctic to a US audience as wilderness and within a handsome album of carefully framed photographs also marked the Antarctic as an object of US national concern, and possibly of ownership. This tension between the documentary features of photography and its instrumentality to ownership, both private and national, continues to generate visual meaning in Antarctica.

Photography in the Antarctic has since the Heroic Age of Antarctic exploration (1895–1917) been the most overexposed of visual technologies, through its mediation bringing into view a territory that before had been represented only in written narrative and paintings and sketches—forms that did not circulate as widely as photographs (Glasberg 2008). Yet, for all its popularizing, and indeed because the medium of photography became so ubiquitous and powerful, its significance in the creation of Antarctica as a place is often subsumed in more technical discourse or aesthetic judgment. As documentary proof of national territorial claims, knowledge-gathering and mapping, and as advertising/promotion, and even as an aesthetic practice in itself, photography has played multiple roles in the creation of Antarctic place even before 1911–12, when both Norwegian and British explorers photographed their arrivals at the South Pole.[3] This chapter builds on the centrality of Antarctic photography to the creation of Antarctic place in order to make a claim about how a particular US landscape aesthetic has emerged within and through the contemporary era's breakdown of national borders and of rising awareness of environmental crisis. The 2008 series "Events Ashore" by US photographer An-My Lê exemplifies this postnational, post-ecological view of the Antarctic. Working within and against the powerful imperial notion of Antarctica as a tabula rasa and the landscape tradition that nurtured Antarctica's blankness, Lê presents Antarctica as a "non-place," produced within a network of US global militarization that questions assumptions about the impact and future of national and corporate presence on the continent.

An Inviting Blankness

But before landing at Lê's "Events Ashore," I need to further expose the history of Antarctic landscape photography in the work of Porter and Herbert G. Ponting (1870–1935), the celebrated British photographer of the Heroic Age whose influence and example extends to Porter and to Lê. Ponting came to Antarctica as an experienced and well-traveled photographer of places far from the United States, where he lived, or Britain, where he worked. But the Antarctic presented specific challenges due to its temperatures, harsh weather, monotones, and lack of more traditional flora and fauna. Ponting's well-earned reputation for photographing remote and little-seen places was based on a traditional landscape aesthetic that in many ways unraveled in Antarctica.[4]

Ponting, the official "camera artist" of the 1910–12 Scott Expedition, struggled to address the relative lack of classical perspective of horizon and scale (Figure 5.1). Initially disappointed at not being among those chosen for final trek to plant the British flag at the South Pole, he later rationalized that there would have been nothing of interest on the polar plateau to photograph. Thus, Ponting chose sites and objects close to the shore, the hut, and to human activity—framing strategies that created variations on classical perspective, with an icy particularity. Ponting helped establish what is now recognized as the Heroic Age aesthetic characterized by dramatic juxtapositions either between human figures

Figure 5.1   Herbert Ponting, "The Castle Berg with dog sledge." With permission from the Royal Geographical Society (with IBG).

(or those of the built environment) and icescapes, or within inherently dynamic mountains or crevasses. The human-scale figures and the encompassing icescapes only become meaningful in relation to one another. Without human figures, the unfamiliar environment would escape the particularity of place. This paradox of encounter—between an indifferent ice and humans intent on colonizing the uninhabitable—fueled the Heroic Age aesthetic. It has left deep traces of style, object choice, and perspective on Antarctic representational history, not the least of which is the powerful imperial imaginary of Antarctica as a tabula rasa, or a pristine, untouched, terrain that, as the translation blank slate would suggest, invited marking. The kind of blankness produced by juxtaposing human figures on the ice or staging perspective with the lens is a "filled in" kind of blankness. It is not unlike the blankness of early European maps that designated the southern continent as Terrae Australis Incognita in words written boldly across the map vellum. The declaration and naming of the territory as unknown was paradoxically a form of filling it in, and of knowing it.

But the type of blankness projected onto or produced through a visual engagement with Antarctica's terrain changes depending on who produced the image. For Ponting, the inviting blankness of Antarctica seemed almost formalist, an aesthetic challenge to create a recognizable scape from such impoverished materials. That his photos were part of an imperial expedition to claim Antarctica and the South Pole underscores the role of cultural concepts in the construction of empire. Ponting's landscapes were more than attempts to fill in the blank of Antarctica with familiar gestures to romantic sublimity: They were claims on the territory created by the camera's eye as much as by the juridical intentions of the British.

Porter is also a national photographer. But the United States played no role in the Heroic Age in Antarctica, and even through later substantial military involvement beginning with R. E. Byrd's 1929 flight over the South Pole and his establishment of "Little America" in 1932, a navy base on the ice of the Ross Sea, and the subsequent greater efforts of "Operation High Jump" (1948) and "Operation Deepfreeze" (1956), has never lodged an official claim to Antarctic territory. So Porter's extension to Antarctica of a monumentalized "desert" blankness directly related to the Yosemite Sierra Club images as an object of not empire, but conservation, marked a very different kind of "filling in" of the blank of Antarctica. "Deception Island" depicts a desert to be conserved, to be kept empty and devoid, paradoxically filled with its own desertedness. But Porter had in Antarctica to choose his landscapes carefully. Sites such as the Dry Valleys that comprised the 2% of unglaciated Antarctica and the more conventionally dramatic mountain ranges and shores leant themselves more directly to the kind of images associated with a sublime nature worth conserving.

Porter was flown to various locations as a guest of the NSF's Artists and Writers Program, keeping extensive notes on the places he visited. He was one of the first nonscience support people to see the South Pole. Although Porter wrote about his trip to the pole, he reveals ideological investments in refusing to photograph the South Pole—a place where within the Porter aesthetic there would be nothing to see but the significant evidence of US colonization in the elaborate, if isolated, built environment. The pole Porter arrived at in 1978 could hardly be considered "empty." It featured a geodesic dome full of supplies and meat locker like dwellings for the personnel. A bright red "milk carton" observatory on stilts and other smaller buildings, a groomed runway, a striped ceremonial pole marker, and various sheds, tents, and areas for vehicles and supplies constituted the scattered settlement. Unable to aestheticize these bare marks of colonization, Porter's conservation ethic in Antarctica leaves visually blank the pages where the South Pole Station or the polar plateau might have stood: It was the wrong kind of blank, an unserviceable lack of visual drama and significance best left to words and traditional narrative. In 1978, Porter empties the Antarctic of all national competition and any trace of the Heroic Age, using aerial and remote technologies of viewing to blank out the landscape aesthetically connected to the US West and strategically side-stepping marks of human intrusion. In so doing, Porter blanks out as well Antarctica's imperial history and so performs a new form of imperialism more suited to the US non-claimancy stance.

"Post–Heroic" Age Photography

Contemporary photographers, including professionals, tourists, and workers, engage Antarctica fully aware of its short history of visualization. Ponting's black and white juxtapositions of ice and human figures created a powerful aesthetics characterized by the trace of the body: the footprint, the track in ice, and the human figure itself. The Heroic aesthetic sees men, ice, materials, and animals all within a range of objects and relations that coordinate toward occupation and claim. Despite and against their own fragility or marginality to the ice, these marks march toward a future of increased levels of habitation: more marks, and more men. Yet the present produced by these marks of culture on the ice can never attain the "aura" of first arrivals (questionably) documented by Heroic Age photographers. The illusory nearness of the original heroic achievements—due in part to the "thinness" of Antarctic human history that as a result overemphasizes the heroic deeds, the ability of ice to preserve for many years the marks of human activity, and the overwhelming durability and iconicity of the barely colonized ice itself—has resulted in a pervasive fascination and even nostalgia for a former heroic landscape and narratives of explorers' hardship and suffering. Because contemporary observers cannot in reality approach the state of the original heroes that nevertheless seem so tantalizingly available in those vast and icy scapes, the post-Heroic in Antarctica is expressed through a critique laden with the desire for the past's traces – a topic discussed in terms of narrative in Chapter 2.

The desire for a heroic unmarked Antarctica and the post-Heroic reality of human inhabitation collide in An-My Lê's photographs taken at the South Pole. Lê's response to the history of blankness at the South Pole is to stare it down. Lê disrupts the aestheticization of Antarctic wilderness begun with the Heroic Age and instead links Antarctica to sites of US militarization around the globe. Lê inherits an iconography from Ponting and applies it to a US Western subject through Porter, but focuses not on the "empty" deserts of untouched wilderness, or the wilderness fantasy, but rather on the traces of human presence mystified in the Porter images. For Lê, the

Figure 5.2   An-My Lê, "Fuel Storage McMurdo." With permission from An-My Lê.

detritus of a built environment become survivals of the Heroic and later environmentally conscious ages that sit strangely, disturbingly on a contemporary Antarctic of international science.

As plain and as chaotic and unfocused as the dispiriting sprawl it pictures, "Fuel Storage McMurdo" documents US presence at the site of Scott's former base, a location that at first glance hardly seems Antarctic at all (Figure 5.2). What Lê has done to create this new perspective on Antarctica is to simply turn around. At McMurdo Station, instead of photographing the main station from the shore, Lê climbs Observation Hill—named by Scott's men and used as the name implies for orienting, it also is the site of a memorial cross to the dead of that expedition—only to point her lens at the back of base where the less picturesque yet essential support structures familiar throughout the developed world are massed. The resulting anti-heroic landscape is reminiscent of the less iconic images of the Heroic Age huts—the ones that showed the scattered boxes, equipment, and disorderliness of the everyday attempt to inhabit the ice. At McMurdo or "Mac Town," this suburban-style sprawl has become institutionalized and inescapable, a mark no longer of temporariness or even excess, but of the very essence of US presence in Antarctica. And it is not a pretty picture.

Like Ponting, who came to Antarctica already having established a career as a photographer of the far-flung British empire, Lê comes to Antarctica from other global locations. Instead of treating the place, its conditions, and its culture as exceptional, she places Antarctica among her "Events Ashore" series linking US military outposts and bases around the world, including Japan, Australia, Kuwait, Iraq, and California.[5] Lê connects and captures the edges and unnoticed perspectives and juxtapositions of nonplaces or transit zones such as aircraft carriers, "foreign" officer housing, or Antarctica—a place lacking natives or traditional nation states. Le's vision of Antarctica as a site in the global economy emphasizes the posthuman, the corporate, and industrial features of contemporary Antarctica. Refusing the overexposed, official view, Lê shifts away from what we might call an Antarctic exceptionalism. Antarctic exceptionalism, modeled on American exceptionalism, creates a separate sphere for Antarctica. Antarctic exceptionalism abounds in nature channel canards: Antarctica has never known war; it lacks people and history; it lacks political turmoil. Yet, Antarctica is no space of elemental nature, and is only obscured as a list of negatives, or blanks. Keeping Antarctica blank has its costs.

Lê asserts the links to capital flow in "Fuel Storage McMurdo," seeming to have anticipated the headlines about the price of oil. The increased volatility of oil prices has negatively affected planning and funding of the very NSF program that sent Lê in 2008 to photograph these barrels of oil that in the present climate represent a completely fossil-fuel-reliant economy.[6] Oil barrels, full and stored or empty and abandoned, have since the 1930s, when a concerted US effort to colonize Antarctica began, been a stock image of Antarctic stations.[7] Where clearing of tree stumps once marked the passing of wilderness, the fuel storage tanks tell similar tales of the technology and resource needs for human being. The massive effort to create permanent structures in a wilderness is nowhere more dramatic than at the South Pole, where nothing survives but ice, and everything, from toothpicks to neutron detectors, must be flown in from the north at great effort and cost.

The built environment at the South Pole, though relatively circumscribed and brief in duration, is nevertheless deeply marked by US cultural projections. Few know that a Buckminster Fuller geodesic dome, or a "Bucky Ball" had slowly been buried in the ice of the South Pole, after having been planned in the late 1960s and completed in 1973 and shielding inhabitants and supplies until 2006. Lê's image of the sunken dome that once carried Fuller's utopian vision of "spaceship earth" is reminiscent of Tacita Dean's "Fernsehturm" series on modernist architecture (Figure 5.3). Sue Hubbard describes this series as "encapsulat[ing] a lost historic vision and an optimistic

Figure 5.3   An-My Lê, "Abandoned Dome." With permission from An-My Lê.

belief in a now defunct social system."[8] Marks of everyday inhabitation such as flags and footprints and bright tape take on an almost macabre feel in the merciless frame: This is a ruin of a futuristic ideal, now obsolete and trapped in ice. The "deadpan" style of direct framing places the dome and viewer on the same plane, emphasizing the harsh juxtaposition of human and built environment. Lê's photography of the built environment at the South Pole emphasizes the obsolete, the fordist prehistory of heavy machinery and the footprint of what might be distinguished as the "US Heroic Age" of naval operations beginning in 1929 and ending with the development of the IPY and the ATS in 1959. While the US brokered the ATS and its international safekeeping of the territory from claims, the period of the US Heroic Age nevertheless resulted in the US effectively and solely colonizing the pole by building the first permanent station in the 1950s, then the Fuller dome begun in the 1960s, and now the third generation of super-sleek corporate-style station completed in 2008 and almost ostentatiously omitted by Lê.

Instead of the new station, Lê investigates the failed and inaccessible, incomplete, or buried traces of the built environment. Lê works against the coffee-table-book aesthetics—her images are enormous, not meant for containment between covers, no matter how glossy. Her high art and highly technical images negotiate between the landscape aesthetic that broke down at the pole and the more vernacular aesthetic that visitors and inhabitants of the South Pole have been attempting to reestablish. The photography of Emil Schulthess, with its use of fish-eye lenses to reflect in the mirror ball of the ceremonial South Pole, its emphasis on sun dogs and other unusual weather phenomenon, is a form of updated "ponting" (after Ponting's carefully—and to Scott's men intrusively—staged perspectives).[9] The extremophile[10] exceptionalism has through repetition and circulation on not least the Internet become an ironic if normalizing repertoire of representation of the polar plateau. Lê focuses instead on the industrial infrastructure, the chaotic, menacing, and wasted space of an unredeemable blankness very different from both the ever-available blank of imperialism's hope and the desert ideal of Porter's conservation aesthetic.

More significant to Lê than the new station is the lost and discarded, the incoherent epiphenomenal and un-beautiful as in "Storage Berms at the South Pole" (Figure 5.4). Lê labels the photo using the argot of the US base as it has derived from its naval origins. The term "storage berm" requires some unpacking, so to speak. Berm refers to an earthwork or a mound formed as a result of human labor, as in the berm walls of a fortress or of mounded earth in a garden. In these terms, the entwined tracks of military and utopian-domestic are evident. But the South Pole is neither fortress nor garden; and the term berm can help to understand how culture in the form of language and built environment

Figure 5.4   An-My Lê, "Storage Berms at the South Pole." With permission from An-My Lê.

is being reshaped in Antarctica. "The berms" in "Polie" patois refers to the entire storage area on the near perimeter of the central encampment consisting of the Dome (now dismantled), new pole station, and outbuildings. An extensive sprawl of plywood platform covered with all sorts of off-loaded cargo forms an almost city-like network of streets. The berm area is like a warehouse without sides or a roof. The bermed platforms require an intensive regime of shoveling to retain their integrity. Far from the "city center" of science or logistics, the berms are the domain of mostly the lesser skilled workers, who are assigned their maintenance. It is on this shunted yet essential sprawl that Lê chooses to center her image, and turn away from the central pole. If there can be a "backside"[11] to a circumlocated pole, the berms are it.

The image lacks contextualizing features beyond a rudimentary and conventional sectioning into sky, horizon line, and foreground of ice. Lê destabilizes this overbearing classical perspective by mixing elements of motion and stasis. Through the overlay of compositional features, Lê also questions the very usefulness of perspective, of relative motion, tracking, and lighting. Despite the bright sunshine and the classical composition, the viewer hardly knows where to focus within the enormous frame. The perspectively broad and flat and multi-focused approach refutes the unipolar cliché and seems to be saying that there is more here in this basic land–sky polar plateau than is generally considered: The pole becomes in this way more like other sites on earth—and in history—than in the extremophile tradition.

Lê records a built environment that is brutally industrial, menacing, and strangely familiar. While Lê's large format images monumentalize the effort to support scientists in Antarctica, it also points to the growing tension between human-powered or based modes of field work and remote, simulated, or robotic modes of information gathering and data production.[12] The human body may not be as necessary to science as an attachment to heroic history might have it. The glorification of labor and the working-class aspect of the South Pole site gestures complicatedly to the impossibility of a nostalgia for the human body as the primary site for meaning on the ice—even as that body is now a worker in an organization within an infrastructure far removed from the dangers faced by the original heroes. Lê's South Pole is a combination of (post) industrial office park and (de)militarized base. The support of science requires an infrastructure whose implications secures Antarctica within a map of globalization and capital flows at odds with depictions of the ice as pure, empty, or even as heroic and sublime.

Military operations are captured by Lê in "Offload," a similarly structured scene of sky meeting water-foreground taken from an elevated position. Though the location is a seashore, not a frozen polar plateau, and the characters are a military amphibious vehicle landing on the shore, a tank, and an aircraft carrier headed toward the horizon, these elements are in analogous position to those in "Storage Berms at the South Pole." The marines on the beach are anonymous and dwarfed by the scene and machinery from which they emerged and which they drive. Relative movement comes through in the spume clouding the landing vehicle, which is caught between being in the water and sitting on shore. A vaguely malevolent, thick sky hangs over the affair. The shared composition of "Offload" (Figure 5.5) and "Storage Berms at the South Pole" demonstrates Lê's anti-exceptionalist positioning of Antarctica as removed from its own narrow history of extremophile representation. Rather, Lê places Antarctica in a visually and politically homologous position to global-ization of the US military. As Chalmers Johnson has observed, the "US has bases in every continent but Antarctica" (Johnson)—though Lê would suggest a minor correction: The Antarctic is not fully or cleanly demilitarized, even though the ATS remains in effect. As Simon Jenkins recently argued in *The Guardian*, the United States and other nations adhere nominally to the ATS restrictions against drilling, assaying, and other development plans, to the extent that science programs can do whatever they want.[13] That includes within science programs, which no one suggests are anything but serious and valuable, activities and implications that are also nationally strategic. These activities help to aggrandize development and may be refunctioned at a time when the empty frozen laboratory becomes a different kind of blank, when ice core drilling and blasting might be put to other less scientifically benign purposes.

The tiny human figures against a grand background echo Ponting's classic perspective. But in this take, the human figures are diminished toy soldiers rather than anchors of the human. Lê amplifies the potential of Heroic Age sublime to an extreme distortion that would comment back on the limits of the very notion of the sublime to give any real proportion to nature or to humanity. This imbalance was immanent in Ponting. Updating Ponting, Lê exposes the absurdity, the menace, and the doom written in humanity's material tracks on ice, and more fundamentally, in the very embodied and logistical modes of human being and seeing in Antarctica.

Porter's anxious warning about drilling in the ice has returned in the form of concern over global warming and renewed resource pressure worldwide. And yet, artists are less interested in the propaganda techniques of the earlier generations of eco-art.[14] A new style of the eco-visual is emerging in which artists directly survey the damage in the landscape, risking its aestheticization and their own complicity in ruin—yet avoiding the equally problematic evacuation of the human from the concept of wilderness.[15] Lê's complex "deadpan" exposure or muted outrage characterizes the post-Heroic in Antarctic photography as both desiring the Heroic and critiquing its trace; lamenting its tabula rasa clarities while refusing everything imperial that its blankness stands for.

Landscape Photography as War Photography

Lê's post-Heroic affect emerges from her background as a Vietnamese refugee-immigrant to the United States who has photographed soldiers and battlefields in Vietnam and in the United States.[16] The platitude "war ravaged" contains and covers many types of devastations, a language of shock and blankness and extinction that Lê indexes on the surface of the ice itself. The dense and overlaid tracks repeat the mark of the footstep of the Heroic Age, the first arrivals, the lone witnesses to a romantic sublime landscape. The layering of tracks comments on habitation and the passing of human time—the mechanical industrial sublime replaces the former South Pole concept by effacing it. The new pole is both marked by and replaces the idea of an unmarkable, vast, timeless land. The marks in "Storage Berms at the South Pole" are trashed, illegible, untraceable to a singular, heroic agent. Rather, these tracks violently blank out Ponting's and Porter's blanking and connect rather than separate the foundations of landscape photography and war or battle photography. This latest form of blanking provides the central argument of Naomi Klein's *The Shock Doctrine: The Rise of Disaster Capitalism.*[17] In describing how free market economic policies beginning after WWII have exploited natural disasters to promote profit as well as policy change to insure future profits, Klein describes a blankness that is unlike the nineteenth-century imperial blankness of the map, whose white spaces were created only to be filled in.

Klein's first chapter, "Blank is Beautiful," describes the aftermath of Hurricane Katrina in New Orleans as an example of the way a natural disaster became an opportunity to politically remake the city to benefit capital, but not necessarily the displaced inhabitants of the ruined city. Klein agues that disaster capitalism not only exploits natural disaster, but produces disaster—through the techniques of shock therapy on the level of individuals, and through the kind of defense policies of "shock and awe" at the center of the US-led occupation of Iraq. Shock and awe were designed to radically undo Iraqi society so that occupying forces might rebuild on terms favorable to their interests. While the blankness in "Storage Berms at the South Pole" is not reducible to the evacuating violence that Klein analyzes, a violent process of marking and overwriting is at play in the composition and subject of Lê's photograph. This new kind of footprint, unlike the original Heroic Age man-hauled and dog-sledded tracks that indexed human–animal time frames and scales and singular if nationalistic effort, is a demonstration for a disembodied, many-limbed, and non-teleological global capital. The South Pole is in many ways a perfect, endless arrival point for a global capital flow, whose very maintenance of the new South Pole requires more attention than any past arrival or claim. The South Pole becomes claimable—beyond the limits of biology or the ecosystem, and beyond the restraints of any treaty limiting national or corporate profit—only within this new economy of extreme capitalism. The contemporary South Pole is a ready-made desert for "shock doctrine" economics.

Working within globalizing capital and big science in Antarctica is a small but persistent history within the US NSF: the Artists and Writers Program (AAWP). Beginning as early as 1957, it has sent visual, sound, literary, and other artists including both Porter and Lê to the ice.[18] Under the aegis and even encouragement of the NSF, a rich record now exists of human presence in the Antarctic that would never have been possible if not for the underfunded program. Yet, paradoxes similar to those of the Sierra Club's promotion/conservancy of Yosemite emerge: national interest in a nonnational project, promotion at the cost of the pristine, obeisance to the state as a factor in its critique. The AAWP has managed to balance the requirements of the state and the need to support science with an allowance for artistic freedom. Although the populations of the southern hemispheric states like Australia and New Zealand incorporate Antarctica much more intimately into their national cultures, the size and quality of the US program has produced a US-authored commentary disproportionate to US cultural involvement in Antarctica. We could even argue that the program keeps the United States in the game culturally and in the important area of geopolitics and the media, or that the AAWP is the cultural wing of a neoimperial project in Antarctica, one that operates beyond the need to legally claim territory, and that surpasses the sundry claims of other nations no matter their histories and worthiness. The cultural imperialism of the United States is promoted, expressed, and yet also countermanded by the NSF AAWP. No matter what kind of art is produced—children's books or Lê's "high" art depicting US militarism—the state cannot fail to benefit.[19]

Can the arts create a bulwark against capital, even as the arts themselves are a feature of the "consortia" that Porter was so concerned about in 1978? We are back to the Sierra Club perplexity: Does national art necessarily presage—and even alibi—industrial development? Lê's post-Heroic, post-ecological wasteland at the South Pole suggests another role for the constructed tabula rasa: that blanking the object may help preserve it from being filled in by processes of resource extraction, whether occurring through mining, science, war, or art.

From Commodity to "Living Ice"

Commodifying ice has been a foundational feature of Antarctica's becoming, as this chapter has demonstrated. Because artists less directly contribute to profit, they have been ideally placed to create value from ice. Indeed, photography of Antarctica comprises a viable market, whether produced by professional artists or by the many "disaster tourists" who travel to photograph the melting ice. For the time being, photography remains a more successful form of commodification than, for example, attempts to tow icebergs for fresh water to melt and bottle. Lê's capture of the process of melting and the invisible "sea" of air into which it melts—or more technically, sublimates—is a potential that produces ice as material; not commodity in the traditional sense, but neither a free-floating symbol to be commodified.

The most important feature of Lê's "Storage Berms at the South Pole" shifts in this materialist reading from the composition, frame and foreground tracks discussed above to the air in between the lens and the landscape. The cloud may be a trail of exhaust from the vehicles; it may be ice crystals kicked up into the air by the friction of tire treads across the surface of ice; it may be a result of windy conditions. Most likely, it is a combination of all these factors. Through Lê's wide-open lens and extended shutter dilation, the miasmic colloidal cloud of ice becomes visible as the very air we breathe, the medium through which Antarctic ice is perceived. This is an astonishing crystallization or visualization of ice itself that in its implication goes beyond the surface of the ice, the location of the South Pole, or the edges of the frame. The molecularized, crystal cloud of ice is not only the medium through which we perceive the scene, it is also the air we breathe. It is being breathed into the lungs of the workers driving the vehicles kicking up the crystals into the blowing wind to be aspirated and respirated; melted and refrozen—a continual processing on a cellular level, an assemblage of living ice. Lê reterritorializes ice, not as a symbol of global connection or the human edict to struggle and survive. Antarctic ice is an assemblage, not an object to be appreciated, exploited, or rescued/preserved; not a blank of projection or futurity but rather a crystal instance, a material becoming of the ongoing present of permanent war, and of the breaking up and redistribution of the continent of ice within global envionmental climate change.

It is worth it at this point to reiterate the cultural illegibility of Antarctica, the colony at the end of the world, in terms of the intellectual projects such as national studies, postcolonial studies, or even postnational studies as they have developed to describe the organizations of the world. The actors involved—the resistant territory, the TNCs, the ATS, the NGOs, and the representations circulated without and beyond the site or object of study—all defy traditional definitions of and relations of power. The conceptual tools to describe the state of non-states or trans-states such as Antarctica need to go beyond the existing rubrics of the national, the colonial, or the global/local. The biopolitical management of populations, images, media, and the ice itself cannot be understood through national models.

Despite utopian and programmatic efforts to manage Antarctica within and against international legal and scientific accords, the global commons and World Park postcolonial concepts and strategies for Antarctica are not likely to take place. Nicholas Johnson, the dyspeptic former South Pole computer tech whose 2006 memoir *Big Dead Place: Inside the Strange and Menacing World of Antarctica* was discussed in the previous chapter as a prime example of a post-Heroic affect on ice. As Johnson's cover image of living room furniture arranged out on a heavily-tracked ice plain and empty horizon, the suburbs have become Antarctica; and Antarctica is an extreme suburb. Suburbs are extensions of civic centers that within the new suburban geographies have disrupted or involuted the center–periphery model. Suburbs are the edges that have become new centers and in this, the suburbs now interact independently of their former centers and in turn recreate and redefine that center. The South Pole has become an office park!

What kind of new citizens do the suburbs produce; and how does understanding US Antarctica as suburban offer new ways to think about biopolitical affects among people, institutions, and the material of place? In the case of Antarctica as suburb, I go beyond the post-Heroic lament and the suburban masculine rec room antics and rage of Johnson and the frustrated inhabitants of the neoliberal workplace. I instead return to an earlier product of the suburbs, the artist Robert Smithson. Since his death in 1973, Smithson's reputation has only grown, resting most on his earthworks. Earthworks deterritorialized art from the gallery and rerouted it back to the landscape, in the process redefining both art and territory. In his many sketches and drawings, Smithson integrates and involutes the surface of geography and the depths of geology to produce landscape that is both situated in time and space and yet able to travel across scales of time and space. The suburbs of New Jersey for him were layered with deep geologic time. The rocks and other artifacts that he sometimes chipped off and dug for were on a par with the sites on the turnpike, the features of his car dashboard, and the memories of his suburban upbring-ing and imagination ("Monuments of Passaic" in *Collected Writings*). Authors of what we might call earth or material fiction, such as Poe and Lovecraft, as well as Sci Fi fiction and film, as well as space travel fascinated Smithson throughout his career and formed the basis of his concept pieces on the Museum of Natural history, the planetarium, and his earthworks. For Smithson, it all came down to the mutability and connection of material. He sculpted with and through mediums themselves. He sculpted with affect.

"Proposal for a Monument in Antarctica" (1969) was discovered in a storage container of Smithson's estate in 2001 (Sobieszek 145) (Figure 5.5). The work on paper, which had earlier been catalogued under a more generic title "Sci Fi landscape," acquired its more specific title. The image depicts a scene of contact with a shoreline jaggedly cutting or splitting the page. On the side of water, indus-trial steam ships are moored. On the other side of the splice, tiny human figures haul ship ropes in unison. A cantilevered stacked cube arrangement—the monument of the title—was drawn by Smithson on another source, cut out and superimposed into the collage. The tiny rope pullers and overseers were hand drawn and give the piece the appearance of an architectural model. Other elements of the col-lage including the moored ships, the strip of horizon at the top of the frame, and most likely the plane of blank ice are found magazine images.[20] The entire collage was photostated, a process of duplica-tion common in business offices of the period. Thus "Proposal for a Monument in Antarctica" is a duplicate set of images, one positive and one negative, one original and one copy. Which is original and which the copy is not only unclear but the question of sequence must be considered in relation to the un-originality of the other elements of the collage that Smithson has put into contact.

Figure 5.5   Robert Smithson, "Proposal for a Monument in Antarctica," 1969—positive image of a two-image set. Art © Estate of Robert Smithson/Licensed by VAGA, New York, NY.

I detail Smithson's strategically jumbled (even the shadows are reversed and misaligned, a parody of the cadastral survey as well as another layer of photographic manipulation) and even violent construction process to emphasize the tension between control of the landscape and the vitality or unpredictability of the materials of that landscape. The doubled arrival of industry and public art in Antarctica emphasizes the ice as a resource of primitive accumulation. The title's juxtaposition of the future of "proposal" and past of "monument" as well as more subtle features of the landscape such as the incoherent shadowing suggestive of multiple suns place "Proposal for a Monument in Antarctica" within Smithson's career-long investigation of a material world that is anything but servile, inert, blank, or ready to be turned into resource—even for a sculptor.

Smithson extends the postcolonial critique of occupation and resource extraction through traditional artistic practices of sculpting stone or metal that mirror geologic processes to the ice. His quarry (so to speak) is the ice itself as the crystal, the basic structure of ice. Smithson produced extensively around the crystal in essays such as the "Crystal Land" and his choice of earth materials, rock, glass, shells, salt, and processed forms such as asphalt and mirrors are all mutations of basic crystal structure. The monument itself, a composition of ice meant to memorialize a continent already composed of ice is surely Smithson's joke on the self-serving quality of monumentalization so often commissioned by private industry. As an igloo-like structure that cannot be actually used to preserve life, "Monument" also alludes to industrial disregard for the peoples and lands on which their profits rely. But the "people" Smithson speaks of and for are not mythologized "Others" of prehistory or predevelopment who must be either saved—or included—in capital development. He does not project a species of native onto the land as a foothold for intervention. The connections Smithson extends onto the ice lead back in layers and trails of ink, light, and earth to his own origins in suburban New Jersey.

The crystal structure had an analog in the practical furnishings of Smithson's car, the very one he was driving on the highways of New Jersey. The graduated flight of white tabular buttons on his car radio formed the perfect structure that conjured a designed universe through which he traveled and which in turn enveloped his body. For Smithson, this panel of white button-tabs morphologically echo tabular icebergs, pyramidal (or igloo-like) constructions, and basic salt crystals; these associations create a shore of contact or sensorial switch point into suburban, subatomic, substructural dimensions of the universe. In this affective–materialist multidimensionality, the radio waves summoned and directed through the car antenna, Smithson's fingers on the radio tabs, his thoughts and memories and sensations, the crystal structure of molecules, the asphalt of molten, crushed rock and shell and tar of the highway, the roads, the flowing of water carrying the ships, shore, the paper, and the light of the spinning sun as the irradiation within the photostat machine are all structurally related in a material flow. The dual temporalities of the future-oriented "proposal" and the backward-looking "monument" extend and exchange continuous and merging roads to pole that are material and virtual. It travels back to the radio tower of Little America, through the already obsolete radio broadcast to "Keep watch on the skies" that concludes the 1951 film *The Thing from Another Planet*, through the obsolete futurity of the Buckminster Fuller dome design of the US South Pole (1972–2010), and into suburban basements full of specimens collected by amateur geologists like Smithson (*Collected Writings*).

Smithson's geologic reterritorialization and sci fi landscapes return ice to the experience of the extreme suburbs, or suburbs on ice.

Le Guin's foundational undermining of the Heroic Age of Antarctic exploration in "Sur" has done more than open Antarctica's history to feminist, postcolonial, environmentalist, and posthuman speculation. It has also demonstrated the art of exploration itself. One of the South American women on the expedition to discover the South pole remains at base camp to work on sculptures. She uses the only available medium, what the narrator calls the "living ice" (264). Paralleling the paradox of leaving a readable historical trail that refuses the reification effects of history (discussed in Chapter 2), the artist cannot return to the northern world to display her creations. That is "the penalty for carving in water" (265) or of being in time with a place rather than with the projected futurity of the historical.

"Carving in water" is also the problem of the place of art in Antarctica. Ponting made a living showing his photographs and films of the Antarctic. Scott, once reviled for his sensitive, artistic temperament unsuited to quasi-military leadership is now appreciated for those very qualities; Edward A. Wilson's watercolors of atmospheric phenomenon have become art objects. These few examples suggest that the relation of art, exploration, and science has become institutionalized with the growth of national artist and writer support programs within national science foundations. Putting aside the question of the art itself, what is important is the institutional route of control through national science foundations, including most significantly the NSF, British Antarctic Survey, and New Zealand Antarctic Program. Lê and the artists discussed in the next chapter fold into their art practices an awareness of their dependent condition within scientific management. Still, the bulk of Antarctic art is not supported or funded directly. Increasingly, it is being self-funded or more usually, produced on Antarctic bases by workers paid to do other jobs. So much more work needs to be done on Antarctic art, especially in relation to methods and practices of science. The next chapter on feminist environmental art begins to do just that.

Chapter 6

Sculpting in Ice: Climate Change and Affective Data

> *Discovery, or uncovering, has to do with what already exists, actually or virtually; it was therefore certain to happen sooner or later.*
>
> Gilles Deleuze, *Bergsonism*

Polar explorers, primarily known for symbolic acts of imperial derring-do, are reliable and safe, nonconfrontational heroes for confusing times.[1] Perhaps, this was all quite clear to Prince Harry of Britain, who joined a February 2011 trek to the North Pole as a member of "Walking with the Wounded," a group raising money for injured veterans. Once the mythic zones of European heroism, the poles are today more associated with concerns over native rights, oil company drilling and privatization, the destruction of habitats and species, and most of all climate change. So the Prince in bright red survival gear posing (for *GQ* and other fashion venues) is only the most recent recuperation of myths of British heroic martyrdom—John Franklin's lost north polar expedition of 1845 and Robert Scott's 1912 disaster at the South Pole.[2]

I am not really interested in flogging the Prince's adventure on ice as a remilitarization of a newly contested region. The desire to discover the ends of the earth and to make use of them had been predicted in literature and scientific imagining long before men finally arrived at the ends of the earth. It was, in Gilles Deleuze's protest against imperial geo-teleology, "certain to happen, sooner or later," and surely it is still happening, on and on and by many different types of people for a myriad of causes and beliefs. Preceding Prince Harry's latest remasculinization and following in the footsteps of the Heroic Age of Polar Exploration and the 1959 establishment of an international ATS, women, non-Europeans, southern hemispherics, people of color have performed physical survival in an extreme environment.[3] Whether by direct representatives of governments, private adventurers, tourists, or more likely, associates of national science programs, polar ice has become evermore filled-in, storied.

Ice has become a sublime incitement. If everyone now travels to the poles, everyone can be a hero, too. The more people trek to the poles, the more they empty the ice of its own materiality or liveliness. They leave footprints, tell stories and take pictures, build stations, and change a landscape from a self-contained and perhaps incomprehensible one into a metaphor of human survival, a trace of past failure for which the only point seems to be to have arrived and to have survived. And yet, most people travel to Antarctica not only to rediscover the traces of human history. They go also to encounter the ice directly. I am concerned here to consider this backdrop itself—the ice—not as a proving ground for humanity, not as a homeland as it is for many Arctic natives, not merely as potential-territory. How can the ice be discovered outside of historical determination? What would ice become on its own?

While the draw to approach the polar regions and to pose and retread heroic histories, to remilitarize and reassert imperial intentions of human time has not ceased, a new feature has emerged. The ice itself is under threat. It is no longer the enemy or the implacable outside force to be conquered. Nor is it an uncomplicated pure wilderness. It is no longer exactly a blank backdrop for imperial posturing. Rather, the ice is fragile, melting, ever shifting—in need of rescue. The controversy about climate change, and especially the computer-modeled projections of melting Antarctic ice sheets and inundated cities that were the centerpiece of Al Gore's 2005 documentary *An Inconvenient Truth*, has shifted Antarctica from a blank backdrop for empire or even a zone of purity to becoming another part of a war-torn and environmentally endangered planet. Especially as a sentimental object of anthropocentric panic—*the ice is melting and we're all going to die!*—Antarctica's ruin now prefigures and announces a more universal ruin. It is the site and source of a new but familiar kind of environmental melancholy and neoliberal sentimentality: Earth must be rescued from the damage humans have caused (Szeman 2008).

This renewed struggle to rescue and survive instigates another round of neoliberal economic negotiation, of technoscientific fixes, and survival and rescue fantasies. This chapter will consider what has happened to Antarctica since the establishment of the ATS and the reconsolidation of Anglo-European power through science-centered management.[4] How is Antarctica's fragility and endangerment connected to the shift from national contestation over sovereignty and territory to a science-only rationale for human presence? Recent work by visual artists Anne Noble, Connie Samaras, and Marina Zurkow all situate Antarctica (and polar ice more generally) as more than a landscape of ruin instrumentalized through sentimental heroism or international science. This essay investigates how these artists work with rather than suppress the paradox of human presence as the frame for approaching Antarctica and foreground the materiality (including gendered histories and embodiments) of the visual production of ice. Through their post-Heroic stances and use of rephotography, digital photographic manipulation, and digital animation, these artists promise a discovery of ice that is not merely an "uncovery" of human history.

A Return to Ice

This chapter's title, "Sculpting in Ice," owes its inspiration to "Sur," the 1981 short story by Ursula Le Guin discussed also in Chapter Two. In the 30 years since its publication, Le Guin's un-discovery of the South Pole on behalf of colonials, women, and all those belonging not to the heroic classes but to the "crew," has anticipated even more than kept pace with developments in the globalization of the polar regions. Chapter 2 developed the implications of Le Guin's insertion of a preemptive Latin American women's discovery of the South Pole in 1911, months before the heroes arrived, specifically in terms of the narratological force of history. The story's final line, "we left no footprints, even" is a postcolonial koan, a riddle for contemplating the absences and enclosures written into the structuring of history, not exempting resistance. Like a good koan, and like a good track, the footprint track of the heroes that Le Guin's women encounter standing "in a row, like cobbler's lasts" bear repeating—or following.[5]

But rather than approaching the footprints as a theme or symbols signifying within linguistic-based history, this chapter takes a new materialist approach: These prints, disconnected from the heroes of exploration who left them, read as a fossil trail of a species itself possibly extinct. The prints, the narrator explains, were created by the winds scouring away all but the pressure-created boot treads. They are in effect man-made outcroppings, a manufactured version of more familiar nunatak (wind-formed ice); even their mode of being speaks of out-of-sequence process, or inconsequence. And as the image of the cobbler's lasts suggests, they are mechanical, contrived, and reminiscent of factory production; foolishly out of place in a terrain that offers so little hope for capital development or for the continuation of existing cultural forms of reproduction. I want to suggest that "Sur" offers more than a critique of the European historical relation to Antarctica that has led to a bad end and that the implications of the undoing of the discovery narrative do not end at the end of the tale, or the bottom of the page as a self-erasing gesture of the very path of critique. Rather, "Sur" offers an embodied and visual approach to ice that promises in this era of climate crisis a return to ice.

The following description from the hoax narrative of the building of the women's base camp illustrates how they followed in the footsteps of the heroes of exploration—a literary, critical act of colonial imitation. But the women also depart from the tracks of history. Metaphorically, as readers of the narratives of exploration and literally, as they sight the marks of the official explorers in the ice, the women depart from the practice of living in huts made entirely of imported wood. Instead, the women build a base of tunnels under the ice, excavating new rooms as they need. Much to the disgust of the all-male crew of their supply ship, the women dig out sleeping quarters of tight "wormholes" lined with hay (50). Further controverting the heroic model of imposing humanity onto the blank of resistant ice, the women call their base a "prairiedog village" (50) and in describing their ongoing acculturation of the built environment to its place, redescribe the polar environment not as the enemy, frozen territory of their heroic models, but as the "living ice."

"Sur" generated a predictable backlash from antifeminist and Eurocentric defenders of the heroic legacy. Less predictable was Le Guin's 1987 essay "Heroes" in which she distinguishes between the quasi-militarism of Shackleton's projection of ice as an external enemy to Scott's resigned acceptance of his responsibility and relation to the power of ice. The ice to Scott was not an externalized threat, but more of an enveloping force. The improvisational changes in plan, in retrospect maligned by every armchair expeditioner, are evidence not of imperial blindness or imposition, but quite the opposite. To Le Guin, Scott gave in to the ice. His logbooks, she notes, exemplify not a scientific or military architecture, but rather an artistic affective relation to the environment. To Le Guin, Scott is more like the artist–explorer in "Sur" who sculpts using the available medium ice, knowing her work can never be displayed off the continent; "that is the price of carving in ice" (51). Sculpting in ice is another koan, but rather than a temporal riddle of marking and erasure, it offers a materialist problem of ice itself. Recent feminist landscape works with the problem of the nonmetaphoric footprint through the materialist problematic of translating the ice into human-centric terms, whether they be historical, environmental, or aesthetic. Recent materialist–environmental artwork does not avoid or refuse complicity with the historical past or with ongoing practices contributing, paradoxically, to the very environmental damage their work in part hopes to fore-fend. The visual art on ice discussed in this chapter pays the price of sculpting in ice in work that unseats the anthropomorphic assumptions of history. But in order to give in to ice's demands, these artists also treat the history of photography, the visual, and the digital as sculptural media subject to the same problematic. What remains after the contemplation of these visual works is the ice itself.

New Zealand-based artist Anne Noble's 2005 "British Petroleum Map" is a "rephotography" of an actual game board-map produced by Shell Oil to commemorate the 1957–8 Hillary–Fuchs's mechanized crossing of the continent, just before the Treaty banned corporations and all economic development (Figure 6.1). Noble has described the explorer-like pleasure she experienced tracking down the game board to a forgotten archive within New Zealand's

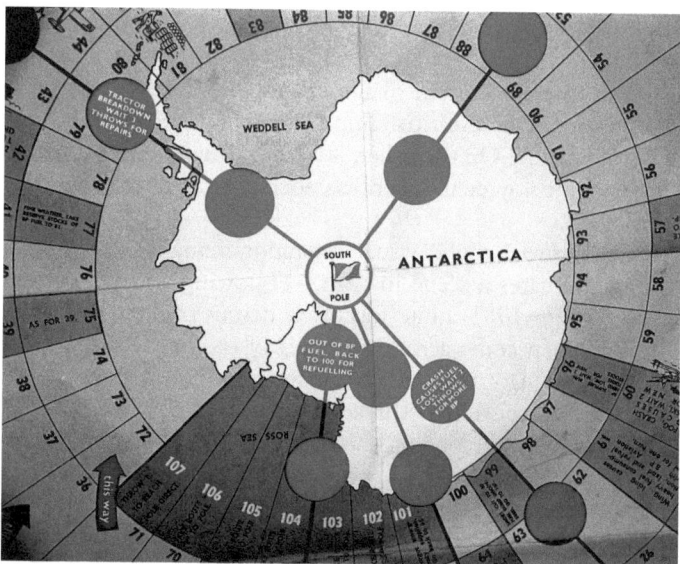

Figure 6.1   Anne Noble, "British Petroleum Map," 2005.

Canterbury Museum, following a tip she had gotten when stationed at New Zealand's Scott Base, where she had been sent as a government-supported artist (Noble 2008). Noble, as an artist/explorer, knowingly following in the footsteps of the heroic explorers, compares rephotography to re-exploration, echoing as well the difference or the space between absence and presence inherent to the indexical and time-dependent modes of photographic reproduction. In the space—which is also a temporal gap as well as the gap between map and territory—between the original and the copy, the possibility of the absent ice itself emerges.

Noble's digital photograph overlays itself point for point like a grid on the original game board. It appears to be an exact copy in the place of the original, which consisted of a cardboard playing surface featuring a highly schematic white and light blue map of Antarctica centered on the South Pole. Bright red dots mark fuel depots and are connected with thick red lines, creating the effect of a orderly, well-supplied, and maintained road connecting the South Pole to the edges of the continent. British Petroleum (BP) trademarks and hand-drawn-looking sketches of airplanes and clusters of oil drums fill in the otherwise blanked-out terrain. One of the red dot turn markers instructs: "Crash causes fuel loss. Wait 3 throws for more BP." The photograph's tight framing of the board has the effect of squeezing out a horizon so that, while resembling a map, the gameboard's cropped horizon instead of offering the conventional mapped "view from above" is instead strangely claustrophobic, more suited to a game of disaster and rescue in a projected icescape. Noble's revision on the one hand releases British imperialism of its responsibility—it is after all just a game! On the other, it releases the potential of national and corporate deployment of heroic associations to once again control and overlay ice.

A similarly melancholic visual retemporalizing of a relic of the heroic past animates a scene in "Sur." Following and yet fictionally preceding the (in 1981) quite forgotten heroes of polar exploration, Le Guin's women come across the trails of the previous attempts to reach the pole. They "discover" the abandoned huts and markers, like an empty oil can securing a "threadbare flag on a bamboo pole" (47). But the many photographs by their precursors of elaborately staged "hero shots," and especially the huts predisposed the women's vision. Expecting a home-like hut nestled against the pristine albeit harsh ice, their widened context instead offers a scarified icescape of seal carcasses, dog turds, and an unkempt hut littered with spilt tins, clogged with ice, its windows broken through. Confronted with tracks of a clearly suspicious and even failed endeavor—not to mention an anthropogenic environmental ruin—most of the women vote to either ignore or rehabilitate the hut; one urges that they burn it down. The scene translates the presentation of exploration's effects in the neat form of heroic narratives of humans struggling to survive on the ice into a visual language. The selective framing of narrative, once disrupted by the women's direct experience, revisions or discovers the heroic past as a ruin in itself; as both dissembling and so ruining the ideal of documentary accuracy, and as ruinous to the icescape.

Rephotography, in its pressuring of the timed distinction of original and copy, takes on greater meaning in a landscape in which questions of origination and historical claim—who got there first?—create the map, if not the territory. Noble's rephotography becomes a mechanism for elaborating, superimposing, extending, and reversing time and space. It is a nonhuman layer of intervention, the meeting of photographic surfaces—the camera lens faces the mechanically drawn game board—to produce the possibility of a nonhuman-generated field. Mechanical reproduction creates the nonhuman elaboration of human-centeredness and begs the question, Is it natural or anthropogenic? This very question instigates and characterizes much of the debate on the causes of glacial melt. The 2010 BP oil spill in the US Gulf of Mexico, a territory far removed from Antarctica, ironically, through its becoming a zone of oil drilling disaster, is now another "copy" of BP's original map of the well-marked and rational road to fuel utopia at the South Pole. The anthropogenic polar landscape has been reproduced, or in geographical terms, migrated. The rest of the earth is subject to becoming Antarctica—or, an extreme and fragile landscape laid waste by fossil fuel capitalism. Thus, Noble's rephotography, which seeks to reanimate the imperial past, turns out to predict the future, too.

Post-Heroic Structures of Science

Widening the frame of Antarctica's production to include postcolonial perspectives however, does not guarantee a less distorting approach to Antarctica. The post-Heroic engagement with Antarctica poses distinct problems for former imperial nations, postcolonial emergent powers, nations who claim a part of Antarctica under the Treaty, and those who do not and yet maintain large science programs and retain rights for the future of Treaty renegotiation. But especially on the level of cultural production, the post-Heroic remains a mode of masculine celebration and lament for which explorers such as Shackleton,

Amundsen, and Scott continue to appeal to an increasingly wide range of people. Antarctic exploration, due to its exceptional cultural features—it is the only continent lacking natives and thus is absent of traditional forms of human exploitation—and its geophysical placement and nature, remains compelling as a way to rescue imperial history of its own ruin. Trapped in all that still-unmelted ice is the hope that heroic bodily endeavor, progress, and technoscientific management may yet prevail against the evidence of the closure of fossil-fuel-based capitalism.

This rescue of Antarctica from the problem of human-generated environmental damage, and from the effects of the migration of pollution to non-populated ice is the new occasion for solving the problem of how to make Antarctica valuable to humanity. Environmental concern has replaced imperial manifest destiny and even international scientific cooperation (as war in another form) as justification for presence in Antarctica. On the way to this neoimperial moment, in between martyred Scott and the military-masculinist parceling up of the continent under the alibi of international science of the ATS, women also arrived in Antarctica.

The problem of gender, or of women more specifically, entering the sphere of Antarctic ice is not, as the bulk of liberal feminist writing suggests, one of a push to equality and a widening of the lens of Antarctic opportunity to include women as workers and policy makers. As Le Guin's parable of the impossible significance of women's prior arrival at the South Pole suggests, the interruption of traditional imperial history to take into consideration new powers and participants shatters that history by leaving it essentially unchanged, and by extension, ever more unchangeable, a permanent mark. A feminism developed within neoliberal frameworks cannot counter the forces that have produced its claims for access and equality. Far from being a problem for Antarctic international management, the advent of women and even ecofeminist principles of environmental protection has fit in all too well. Post-heroism, feminism, and science all cohere under Antarctica's neoliberal management.

Jerri Nielsen's celebrated 1998 "rescue" during mid-winter at the South Pole exemplifies the ultimate coherence of the difference of women's bodies, liberal feminism, and neoliberal scientific management in Antarctica. Nielsen was the doctor assigned to South Pole Station when she discovered her own breast tumor. She self-diagnosed and received chemotherapy treatment through a variety of remote technologies of communication and material transport until a plane could be flown in to retrieve her (Nielsen 1998). Nielsen's rescue echoed earlier heroic embodied struggles for survival and was cast as a US national drama, despite the actual lack of US sovereignty over Antarctica. Despite popular perceptions of the Antarctic as frozen holdover to a male-only environment and thus debilitating to women seeking professional advancement, women have easily followed the footsteps of the heroes—in fact more easily, I would argue, than contemporary men. *Big Dead Place: Inside the Menacing World of Antarctica*, a 2006 memoir by Nicholas Johnson, who worked at the South Pole behind a computer, exemplifies the twinned desire and disgust for a masculine, heroic past.[6] As a participant observer of a little-known work culture in which idealistic professionals sign up in hopes of communing with pristine wilderness and cute penguins only to find themselves trapped on base washing dishes, Johnson's affect of working-class ressentiment makes it clear that Antarctica has become far from the proving ground of masculinity (failed or not, reviled or not). It is rather just another office park in which Johnson seethes at the poor leadership and failed policies of his superiors at Raytheon Corporation, the weapon-maker and science support contractor for the US NSF. But more than critiquing the idea of Antarctica as detached from the global economy, this office park is linked to a chain of US military bases outside the territorial limits of the United States. The post-Heroic takes place in the context of neoimperial restructurings of military and economic relations of states. Antarctica is not only a site of fascination and pity, but also of neoliberal organization and trade: the perfect destination for the anxious consumer of the world.

The heroic rescuer of Antarctica from its abjection and ruin, or his latter-day cultural stand-ins, the tourist/appreciator of Antarctica's "uniqueness," the stunned viewer of its landscape, are no longer viable positions. Again, the complications of a more complete recognition of the demands of an Antarctic landscape being produced through human-generated interference derive from the cultural–legal shifts of a science-based, feminist-friendly system. It goes almost without saying that the post-Heroic occasions and is occasioned by the advent of women working and living in Antarctica, taking part in its contemporary culture. It is no secret that women's gradual arrival in Antarctica was resisted and finally lamented by the males-only club deriving from adventure, military, and science legacies. Certain "OAEs" or Old Antarctic Explorer types (whom Johnson mocks in his memoir) were not silent about their perception that women would spoil the purity of a homosocial work—and play—environment (Graham 1994).

Yet, clinging to the legacy of liberal feminism's overcoming of barriers to female bodies and social organization carries with it its own
set of blindnesses to the ways that feminization is not simply about
women coming to Antarctica or the empowerment of women as a class.
Certainly concern about disruptions within homosocial organizing
preceded women's actual bodies. And women doing the same work
as men in itself is no guarantee of a significant regime change. From
the perspective of military control, science too, as a post-Heroic entity,
bears the mark of feminization. Tensions between "beakers" and support workers—class as work divisions in contemporary US NSF program and its military-era holdovers—attest to a gendering of science as
feminine and support workers as authentic and masculine. Science and
feminism differ in the regard that science attempts to recuperate the
lost heroic through mastery and management. But non-liberal feminism can never be properly heroic—though celebrations of women in
Antarctica like Jerri Nielsen generally fold back to support a heroic mas-
culinism and the state. Thus, from the perspective of all-male Antarctic
utopias, feminization-as-progress is also a form of ruin, a puncturing of
a homosociality that defined and kept stable nationalist and disciplinary structurings of the continent.

Noble's rephotography and its closed-in overview of the game
board's map view from above heightens evermore the sentimental-
izing of Antarctica under two regimes—one of military–industrial
scientific management and the other of rephotography's own feminist
history. As Rey Chow discusses in *The Age of the World Target*, in the
shift to a critical self-referential worldview under postcolonial, feminist, and other "area" studies, the entire world is produced as a target,
as an object of study. And in Antarctica, that object of study is the
territory itself in need of rescue (Chapter 4). The red dots of the BP
game board map as rephotographed from above by Noble are also x's
and o's of the visual violence of the creation of the "world target" that
is extended through the production of Antarctica as endangered. This
world—the one fully released in Noble's rephotographed game board
map—is no longer a territory to explore and stake national claims
over, but a grid of possibility for neoliberal management and miti-
gation that only retrenches corporate–military power. The Western
consumer's anxieties over global destruction and climate change—
fixated always on the image of melting ice and drifting poles—are
merely the vehicle through which this neoliberal fantasy travels and
finds its reason for being.

Like Noble, who understands her intervention on Scott base as in
part a "shadowing" or coming after New Zealand's Antarctic hero

Edmund Hillary, Connie Samaras is also a state-supported artist, but for the United States. While Noble perhaps discovered herself playing an explorer, Samaras has consciously worked with impersonation, disguise, and other consciously feminist techniques to break into secured and surveilled environments. Being a grantee of the NSF's Artists and Writers Program and gaining access to sites in Antarctica unavailable to even the most intrepid adventurers certainly counts as one of her more daring and elaborate schemes for getting into controlled areas (Samaras). Also like Noble, Samaras shadows the logistical movements of science support, as she herself is merely tolerated and more often resented as excess baggage in a science-run territory. Yet, rather than directly building on the scientific logistical foundation, in each of these photographs, Samaras produces views and images that neither science nor fiction has portrayed. In a way, the views Samaras offers simply cannot exist to be seen within the networks that have nonetheless supported her camera and her eye, a post-Heroic positioning that both implicates the camera in the human-generated scenes it depicts—and also frees.

"Dome Interior" (Figure 6.2) employs the most obvious photographic manipulation, using digital technology to produce enantiography—a biomorphic, symmetrical reflection—of the homey red fabrications of the buildings under the dome. In its mirrored view, "Dome Interior" seals in its meaning better than the dome itself, whose geodesic triangular components notoriously tended to frost over and which did a poor job of insulation, not to mention were a constant maintenance problem.[7] But Samaras's symmetrical and artificially brightened photograph returns the dome to its utopian possibilities. It is full, warm, and intact as the real dome never was. That

Figure 6.2 Connie Samaras, "Dome Interior," courtesy of the artist, copyright 2005. From the series V.A.L.I.S. – vast active living intelligence system, archival inject from film.

dome was in fact abandoned to the burying ice and finally dismantled in 2010. Samaras's "Dome Interior," six years after its creation,
has itself accumulated significance, like a non-ruinous ice buildup. It
now points to an elegiac past of the dismantled dome and equally to
the virtual, completed dome of its own photographic materiality. It
is like Le Guin's "Sur," a retroactive, recuperative act of possibility,
cleverly sealed from critique by the need to differentiate itself from
the heroic past it both models and demolishes.

Samaras proposes digital photography as the mode for recuperating the lost possibilities of humanity at the South Pole. This is a vision
that ironically revises the circumstances and possibilities of embodied human sight (and being, since these were the living quarters) in
Antarctica beyond the limits of a difficult and even degraded built
environment. In this sense, Samaras reanimates a Buckminster Fuller
geodesic dome, that in Sue Hubbard's formulation, represents in its
disrepair, "a lost historic vision of an optimistic belief in a now defunct
social system" (Hubbard). "Dome Interior" is a feminist remembering
of a technoscientific masculine failure. Because it was not predicated on
loss or on a specific embodiment (male, female, or other), but rather on
a built environment whose system had a life outside of the human— it
is eminently recuperable for serious or camp effects by feminism.

Samaras titles her series of photographs, video, and written expression of the built environment at the South Pole after science fiction author Phillip K. Dick's 1981 novel VALIS, filling it in with
new associations: V.A.L.I.S. (vast active living intelligence system)
(Samaras 2008; Newhouse 2008). Like Fuller, Dick is associated with
an obsolete futurity, but through the name borrowing, Samaras both
reanimates Dick while offering another way to think of living ice as an
alien being without the connotations of embodied difference or the
standard projections of otherness—in other words, a non-anthropo-
morphized presence on ice. The living ice of Samaras's VALIS estab-
lishes an interactive, fluid field for the elaboration of a possibility not
necessarily predicted by its means of production. Antarctica's ice is no
longer a picture to be taken or an object to be studied, recorded, or
even rescued, but a vital force in the universe.

In fact, given Samaras's facility with impersonation, science fiction, and alien or nonhuman life, "Antennae Field" may very well be
a production of VALIS, or an impersonation of or experimentation
with posthuman seeing (Figure 6.3). Rather than staging a horizon
or frame, Samaras seems to give in to the whiteout and monotony so
often ascribed to Antarctica's icescapes. The bleed into infinity of the
photograph's frame and the vaguely downward-angled point of view

emphasize the helplessness of the viewer, not the power of the photoscopic gaze. Although the view is from above the surface of the earth, it is far from a majestic or authoritative standard view from above that Chow argued creates the "world target." The angle is more of a hovering, a compromised, dizzying nonauthority. The disembodied and disoriented viewer searches for signs in the wind-patterned ice.

Shadows from these nunatak, rather than aid with a sense of perspective missing on the vast ice plateau, further degrade any attempt at grounding amid their multitude. However, two imposed shadows emerge from the field. They are the antennae of the title, shadow/light lines that fix the field around them. Or do they? What has created these shadows? A machine, an animal, or the photographer herself? Perhaps, thinking of how biologists strap cameras to the skulls of penguins to capture their perspective moving under the ice where humans cannot follow, this view records an alien perspective? The shadows too pose questions: Are they substance in themselves or merely evidence of presence? In their insubstantiality shadows further question presence, while antennae typically connote an insect or alien intelligence, a cultural connotation complicated by the less obvious

Figure 6.3 Connie Samaras, "South Pole Antennae Field," courtesy of the artist, copyright 2005. From the series V.A.L.I.S. – vast active living intelligence system, archival inject from film.

fact that humans are aliens to the ice, where the material facts and operations of being and especially seeing have no guarantees.

Antarctic whiteout—the condition of too much light paradoxically producing disorientation and even blindness—is perhaps the most common cultural signifier for Antarctica. Yet, Samaras complicates what has become an almost comfortable truism about the difficulty of human seeing in Antarctica. Instead of the cold comfort of an indifferent or even menacing icescape whose projection has nevertheless allowed humans their precarious purchase on ice, Samaras offers a more intricate, involved sense of how human optics are shifting in themselves (as biological being has always been open to shifts) and is changing Antarctica indelibly. There is no anthropocentric, self-serving lament in Samaras's Antarctica, no letting humanity off the hook for its interference—so often ruinous—in and of this icescape.

Samaras inserts the shadow of a presence that is not, or has never been, or cannot be proven to exist. She imprints with light to capture the ghostly or the alien or the nonexisting, the unseeable: the virtual. She uses photography and landscape to undo the very history of landscape photography as a seeable, light-drenched representation. Sometimes light is the problem, not the means of creating an image. Light gets in the way, is wrong, is too much (whiteout); sometimes it erases. Seeing in Samaras's work dramatizes the unfitness of a practice of seeing borrowed from reading signs. Among the limits of sight-reading as detection is that its practices always lead back to the site of human seeing: What is it—an alien! Who is that shadow? Oh, it is me. Samaras's landscape implicates the viewer: You cannot simply look without doing, feeling, becoming part of a problem defined by the creation of the landscape. Binaries of science/fiction, absence/presence are put under impossible pressure. Samaras seems to be inventing a view, some data, an idea that is beyond science to produce. Samaras threatens to plunge the comfortable viewer, to crash-land into what Deleuze calls the "non-human landscape of nature" and what I might consider here, the living ice (Bogues 2004: 66).

The artists I look at take feminist positions, standpoints, or histories and ask posthuman questions to get at "just ice." What they do is to offer a possible solution of Deleuze's initiating paradox of discovery as predicting and predictable (and thus endlessly repeatable), following Liz Grosz's formulation of a material, nonhuman universe of potential difference: "The [material] universe has ... the possibility of being otherwise not because life recognizes it as such but because life can exist only because of the simultaneity of the past with the present that matter affords it" (150). Following this philosophical trace from

Deleuze to Grosz, the remote and extreme ice stands in for the possibilities of an expansive universe that cannot be thought, seen, felt, or contained from a human-centered moment and that yet enables the possibility of continuity. This problem of nonhuman sequence is the problem of Le Guin's disconnected footprints, Noble's temporal–visual lag between original and copy, and Samaras's impossible seeing. It is the promise of living ice.

To understand the promise of living ice, I must return to the material structure of science (including environmental science) as the center of Antarctica's contemporary culture, a problem that has been developed in each chapter of this book. The Treaty's instantiation of science as the *raison d'être* for human intervention, while providing a course for national conflict over territory and resources, has done little to positively redirect those classic territorial imperial or economic motives that were so much a part of the impulse toward human engagement with the continent. In fact, the shift to a science-only alibi for human presence in Antarctica greatly distorts all other aspects of the former heroic structuring of Antarctica. Putting aside the patent value of the data derived through Antarctica's scientific management and the geopolitical and relative environmental protections ensured under the Treaty, it must nevertheless be said that an all-encompassing yet motiveless scientific rationale has allowed empire and competition to thrive while seriously distorting or foreclosing fuller cultural engagement with Antarctica (Elzinga 1993; Turchetti 2008).

Science is itself becoming transformed by its role in Antarctica through its shift from a strictly territorial imperative to assaying for bio or mineral resources and to data-gathering as a resource as its new prime activities. Many opinions are on the contrary: the ATS's promotion of scientific internationalism works not by protecting Antarctica's environment from development or harm, but through only one mapped (historic-spatialized) projection of Antarctica, a particular mapping of Antarctica for the benefit of neoimperial powers. Opening up the map of Antarctica outside the control of science-culture, Samaras's—and Marina Zurkow's—virtual landscapes ask what would it mean to think, see, experience (or *not* experience) Antarctica from the point of view of lively ice.

Samaras's "Night Divide and Contrails" challenges official history and science by looking to alternative data or signs of life at the South Pole (Figure 6.4). The contrails of the title are tracks left by airplanes and thus evidence of human presence—guided flights across the atmosphere—that cannot be traced in conventional ways. Samaras produces signs in a new landscape—the sky—and invites interpretation of this

Figure 6.4   Connie Samaras, "Night Divide and Contrail Pollution," courtesy of the artist, copyright 2005. From the series V.A.L.I.S. – vast active living intelligence system, archival inject from film.

mysterious trace of humanity's effects on Antarctica. But this trace is neither legible nor temporally stable. Rather, the question of the origin and significance of contrails is itself infuriating, antidisciplinary, and paranoid, offering like a good paranormal investigator, too many connections. What are airplanes doing above the poles? Where were they coming from? Or headed? This is part of the unsettling drama of an anthropogenic landscape that Samaras is both discovering and producing, or producing in discovery. Nationalism and science, along with the category of the human, all fade or become displaced by the shadow of the alien that is us. What kind of place is this? What kind of gaze can it support, does it demand? What happens to sight, to the human in this landscape? These questions push beyond the realm of aesthetics and photography, but they do not leave seeing behind. If you can resee Antarctica, you can reorganize the body of the human—and unsettle and reimagine political engagement beyond the bodies of science and nation. That is the science fiction pointed to by Samaras's vision, one in which the land, the human body, the landscape, the camera, and light and darkness/shadow are part and parcel of a way of seeing, an anthropogenic landscape of active change.

I have been presenting the various ways artists create data where there is none or it is somehow controversial or even suppressed in order to counter the post-heroic scientific structuring of Antarctica. But climate controversy aside, there is a level of my interest that is more strictly formal, about how specifically photography can and cannot track the visible. The invention of these contrails—almost invisible tracks in the sky—challenges not only a history of optics, but the tradition of landscape art. Samaras layers a paranoid narrative about alien (or governmental) secret or dangerous presence onto a problem of mechanical reproduction of embodied sight itself. But the paranoia does not stem necessarily from the content of the information or data. Rather, visualizing itself is the source of destabilization. It is always a tracking of vision that, like a map, encodes power dynamics and ideas about terrain and space, but in its very assumptions of being visible, is to be read only by humans. Put another way, Chow's "target" Antarctica must still be legible by the discipline of science—or visible to the eye—in order to be produced as a target. But a secret, illegible, invisible target—the kind produced by Le Guin's counter-mythic women explorers, Noble's undutiful imitation, and Samaras's indelible shadows—promises a non-anthropogenic landscape beyond the merely visible. Of course, that landscape would also have to be outside the scope of photography. More than a grand gesture toward the limits of landscape art traditions, in pressuring the methods of data accumulation and analysis, Samaras implicates her own photograph as another piece of questionable evidence in a regime of scientific management.

The shadow as evidence is also the shadow of a doubt, since it is the mode of seeing itself that has produced this shadow trace, and this questionable proof. This is a description of the paranoid progress of science's data-accumulation of climate within a political context of warming crisis—understood in the Antarctic context as a problem of literally melting substance. If the problem of how to read the signs written on the ice was not difficult enough, the anthropogenic landscape is itself characterized by decay and degeneration. The ice cannot be understood as a substrate or an "opaque plenitude" of materiality (Coole 2010: 10). That material ice so easily converted to symbolic enemy, wilderness, or data field is as we shall further see, a very unsettling and unconventional terrain.

Unlike Noble and Samaras, Zurkow produced her polar landscape without direct experience and without an "original." "Elixir IV" is a 2009 single-channel computer animation of a bottle floating and spinning in an unidentifiable and in fact entirely invented Antarctic sea (Figure 6.5). The bottle slowly bobs among the waves and clouds as it revolves on its axis: an image of calm containment and self-referentiality. Distinctions of containment and porosity, inside and outside, dissolve as they turn the computer screen that is

"Elixir IV"'s "canvas" into a deep mirror, a two-dimensional portal into alternative dimensions. Zurkow calls what animates her bottle "magical thinking," a phrase that evokes snake oil, perfumes, miracle cures—all the obsolete, failed, and scam technologies of healing. Indeed, the animation floats in a chaotic sea of cultural references, all the while containing and reflecting them. Drawn cartoon-like and projected on a screen, "Elixir IV" invokes the 1960s' TV show *I Dream of Jeannie*, cleverly melding the astronaut's capsule dropped into the ocean from space and the genie's bottle of domestic entrapment of a vengeful female spirit. The bottle, with its simultaneously reflective and transparent qualities, is also the mirrored disco ball of the 1970s, a gyroscope, clock, compass, quasi-geodesic dome, a shelter from the polar storm, even an image of the mechanical "mind" that perhaps created this animation. It is the center of a polar sea or maybe an abyss. It might be the "eye" of the climate crisis. "Elixir IV" oscillates between motion and fixity, the graticule projecting from its enlightenment centering and creating, drawing out and forth the anthropogenic landscape—one that never has existed, quite.

If the landscape may never have existed, the creators of landscapes also come into question. Authorship, originality, possession, and responsibility for creation all emanate like lines of latitude and longitude from Zurkow's floating bottle as if it were a homing device dropped or lost by an unknown civilization. But its computer generation is not a means of simply erasing or sidestepping the mark of the human. The bottle contains endlessly looped images of fleeting

Figure 6.5   Marina Zurkow, "Elixir IV" (2009), production still. Courtesy of the artist.

human figures, fragments of Leonardo da Vinci-style perfected human forms diving into internal waters as randomized clouds scud around them. The Da Vinci cartoons are ghostly; they are sketches of originals that never could appear to be seen in this polar scape, even if it had actually existed.

One of the effects of the digital production of a virtual landscape and its presentation as an animated loop is to reflect on the conventional distinction maintained between analog and digital, the mark of presence and the production of images without originals. Zurkow shows how the landscape was always virtual. "Elixir IV"'s graticule is a magic lantern, an obsolete technology reanimated by the digital, an engine for producing form itself. Maps, the laws of physics, animation, projection, the modeling familiar on TV news weather maps—these all imitate life as they also create, or simply, are life. "Elixir IV" contains what the official story of science cannot show and the connections it cannot make. This virtual grid suggests displacement of the grid itself; the satellite on which the data for GIS is dependant has crash-landed in a digitized non-place. This is a digital creation open to degradation, to loss, and to time (Roberts 2000; Yusoff 2007). It is anthropogenic but not ontologically so. And in the possibility of non-anthropogenic origins emerges another spin on the anthropogenic landscape—and on feminism. In offering a polar landscape without originals and without direct experience or human bodies, Zurkow challenges feminist art production based on a "real" body. But more, she challenges an art history of landscape based on land. Elixir IV contains but does not reproduce the actualized icy terrain. Thus, it offers the virtual not as completion or replacement but rather as just ice.

Just Ice

Whiteout, a 2009 thriller set in a contemporary Antarctic base opens with the "thrill" of replacing the traditional male hero with a female US Marshall arrived to solve a series of murders on what is often referred to as a "continent of peace." Post-Treaty Antarctic fiction is full of heroes fighting both the elements as well as corporate and state control. The traditional heroes fight to survive the limitations of their own bodies, and against evil doers, often projected as aliens. Departing from much Antarctic fiction, there are no aliens for the Marshall in *Whiteout* to fight. Instead, after the reveal of the hero-as-female in the opening shower sequence, the story flashes back to a Cold War–era plane crash into the Antarctic, whose discovered remains set off the plot. But this suggestion of geopolitical tension in the internationally shared continent turns out, like the hero's sex change, to be a total red herring. In the end, the problems on the ice have not been caused by geopolitical intrigue, philosophical hand-wringing over purity or oil, or even any invidious distinctions of nationality or sex. Rather, it has been individual greed. And the objects of that desire fill the final shot, glittering with a similar light of the continent's composition: ice. But instead of the vast and unclaimable and unvalued terrain, the plot-producing ice is a handful of diamonds, the false treasure imported through human imposition and history onto the persistently blank and unending ice.

The replacement of ice's fugitive materiality and value with the prismatic artificiality of diamonds suggests that sex and national difference and even history itself, heroic or not, is flattened, thrillerized in turning Antarctica into just another backdrop, a "Big Dead Place," for a diamond caper. But there is also a deadening effect around sexual equality and the advent of women—they change nothing on or off the ice. And this is not what feminism, not even liberal feminism had in mind. Law and order (such as the Treaty) prevail to the extent that the materiality of ice is evacuated. The female Marshall represents the end of feminism, history, and the heroic possibilities for Antarctica. The movie was a complete flop. Maybe audiences know that a cleaned-up Antarctica without aliens and without the promise of disaster—two modes of vitalizing ice through the virtual—is really just another office park at the bottom of the globe. If the explorers of the Heroic Age arriving at the last place on earth ruined a version of imperial history predicated on open territory, scientific management and feminism are despoiling Antarctica (and the poles) as ineluctably as the spread of globalizing capital and the threat of the melting ice of climate change.

The visual pun of that final shot of glittering paradoxically reinforces the problem of humanity in Antarctica: It cannot imagine just ice. Anthropogenic landscapes are quintessentially managed and surveilled landscapes that self-elaborate on and auto-critique their own structuring. To this extent, they are connected to what Naomi Klein has described as "disaster capitalism," whereby environmental and economic crisis become indistinguishable and indeed coproduced. The ubiquitous phrase "economic meltdown" captures the force of this naturalizing push to cohere climate and market systems—the ice is literally melting—within neoliberal management (Chapter 5). The international science regime governing Antarctica, far from being an exception to global unequal economic and military power arrangements, extends and exacerbates those inequalities. In exceptionalizing the continent as a frozen laboratory for science, national science programs have connected it more securely to domination by militarized management. Noble, Samaras, and Zurkow, in their efforts to document anthropogenically produced landscapes, create the evidence of the damage of their own optics. But unlike managed science, these artists do not offer science—or art, for that matter—as a rescuer.

The indirect evidence of warming/melting is at the root of much science debate. The models created by scientists to capture climate change clash within notions of human-time, since climate is always changing. Human timeframes are seemingly fixed as well within the photographic frame, where a sense of change over time is elusive. But rephotography offers another way to unfix photography. So does digital manipulation, shadow play/reframing (and whiteout too). Yet, I am less interested in the photographs as evidence in themselves and more in the ways they point to Deleuze's percept as "the landscape before man, in the absence of man" (Bogues 2004: 66).

The production of evidence of climate change and subsequent calls to address the problem tend to direct attention toward the unseeable "man" everywhere in the landscape and away from the palpable and on-goingly permanent changes on the ground of human presence. This man in the landscape is the very presence that has produced both the scientific evidence of climate and the cultural impetus to manage it. This is the presence Le Guin and later Noble, Samaras, and Zurkow point to without pointing. They produce evidence of effects that emerge from the production of the art in place of a science (and political conditions) that cannot see or read the evidence of damage as such. But the evidence pointed to by these artists is not only or reductively of damage. Rather, it indicates the vitality of the landscape, of ice itself, or of the "landscape before man, in the absence of man" and of the material working of the universe.

Even in the post–Heroic Era, heroic routines and stances continue to provide impetus for geopolitical strategies as well as critiques of their limitations. Climate crisis panic over the ruin of the environment refers back to a heroic agency of rescue from the very circumstances that have produced the need for a rescue from climate crisis in the first place. In the feminist art I have discussed, the impossibility of escaping the multiple materialities of ice interrupts the footsteps of a self-serving referentiality within climate crisis discourse. The feminist art here takes shape within neoliberal state–supported science and instead of protestations of innocence or bitter lament or technoscientific progressiveness or apocalyptic thinking, it considers the difference within science's possible methods and future by inventing positions, views, methods, and data that both builds on and "dis-re-covers" the history of living ice. Anthro-centered technologies and practices including postcolonial and feminist-inspired movements to include non-Europeans and woman actors that carry over into fields and realms, like Antarctica, for which they are not inevitable nor a given, implicate the possibility of a future that is discontinuous from the past circumstances that produce them. Perhaps, in following this discontinuous, at times self-creating and at other times self-erasing path, there is a chance for ice to be just ice.

Epilogue: Becoming Polar

The entire earth becomes polar in Anne Noble's "Bung at the South Pole on a Blow-up Globe." New globes of ocean and ice home in on territories once assumed merely obstacles on the way to somewhere, or as spent ends, nowhere (Figure 7.1). The "Oceanic Turn" in cultural studies gives a new territorial objective to the decentering of humans. Noble uses a child's blow-up globe to simply upturn the conventional perspective on the inhabited globe to reveal its "bunged" South Pole. A bung is a somewhat anachronistic, even salty, nautical term for a plug in a wooden cask that neatly evokes the voyages of

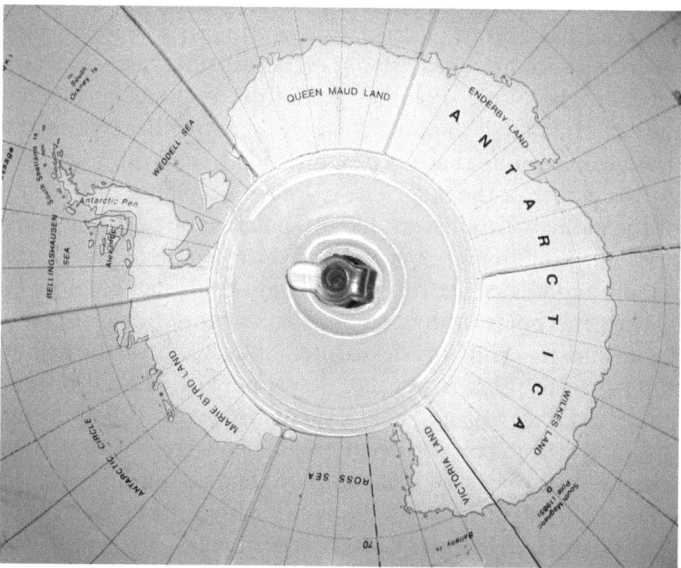

Figure 7.1 Anne Noble, "Bung in Antarctica on a Blow-up Globe," 2005.

discovery and the often sloppy ocean logistics of supply. As a familiar air valve on which the integrity of the blown-up globe rests, the bung signals an understandable panic about ecological risk: it is a precarious stopper for this new "hole at the pole." The bung of Antarctic ice is the repressed, forgotten, and abject key to environmental integrity; it refunctions the imperial pleasure of the map to spectacularize the territory as a possession. Whether a hyper-technical satellite image or as the hokey mass consumerist plastic toy, posthuman maps of Antarctica unsettle the power of even remote, scientific, or culturally "knowing" views of the earth: this earth, and the humanist modes of knowing and producing it is, in another turn on Noble's blow-up toy, also under threat of being blown away.

Climate had functioned much like the oceans as naturalized borders, keeping the poles remote, excessive, and wasted. Climate—and its volatility—has indeed created new poles as a geophysical process and as it has become a sublime discourse in itself; it has also changed what Antarctica means to humans and how humanity attaches itself to the territory of Antarctica. It has been long-coming and immediate; massively distributed and local; overwhelming and indiscernible. "Becoming polar" is the way that places under environmental pressure have been "polarized," or fantasized as pure only to be devastated; simplified only to be ruined by intervention.

Antarctica has become a crucial site for international climate science even as its ice has been calving off at alarming rates. Even more dramatically, the North Pole has gone from impassable to a newly open sea of exploitable shipping routes. Space itself has migrated and that the polar regions that once signified extremity and waste now occupy the center of the map. The poles are becoming more like the rest of the globe. Corporations like BP polarize by pursuing increasingly risky drilling in parts of the world once considered too extreme and remote and wasted, extra. Extreme industrial practices once associated with the poles infiltrate nonpolar territories. The BP oil rig explosion into the Gulf of Mexico demonstrates how the region is "becoming polar"; that is, wasted, desperate, and in need of rescue. Becoming polar signals an ecological switch point. Where formerly humans migrated to the polar regions, it is now the poles—the ice— that is migrating. That ice is moving on its own and moving—that it is emotionally significant, to humanity has constituted the central problematic of *Antarctica as Cultural Critique*.

The impossibility and failure of human being on ice, as opposed to ice's being, is part of the scalar incommensurabilities of the current crisis around how to respond to the climate data steadily amassing. The desire to impose human ways of knowing and being on the material of the earth and the equally strong refusal to fully recognize human inscription and its constituting limits is a problem, which like climate change, has begun with modernity and is only now overtaking human capacities for sensing. It is appropriate, then, to conclude by returning to Antarctica's first fully professionalized photographer, Herbert Ponting, the self-named "camera artist" of the British empire's final expedition to the South Pole in 1910.

"Adelie Penguin and Sled Tracks Crossing, November 1911" depicts a landscape off of Cape Royds (Figure 7.2). It is singular

Figure 7.2 Herbert Ponting, "Adelie Penguin and Sled Tracks Crossing, November 1911." With permission from the Royal Geographic Society (with IBG).

among Ponting's work as an image absent of living or built figures or major scenic features (mountain, shore, ice cave, etc.). Instead, two sets of tracks intersect—one of sled tracks with human footprints and the second an Adelie penguin track—in the center of the ice field, tracing the shape of an X in the otherwise unblemished ice. A distant low-lying landmass to the left and a vaguely seen pressure ridge to the right frame the top edge of the 5 x 7 inch print. Hardly a landscape proper, as the caption suggests, the real interest here is in the foreground shape—the foreshortened, centering X. This perfectly composed image is a play on composition itself, on the (cultural and geophysical) blankness of Antarctica, and on the problem of depicting it. Taken after the departure of the polar party and before Ponting's return to the United Kingdom, the image is pure poise, balancing the disquieting blankness of the icy terrain with a cleverly "found" object. For once, Ponting avoids "ponting," or the careful posing of objects, especially the crew, to create juxtapositions of tone and scale and the implication of dramatic narrative. Instead, Ponting fixes on an abstract scene evacuated of biologic actors and investigates their trace, which become classical lines of perspective crossed in upon themselves, involving and mirroring the landscape in the act of composing it.[1]

The figure that in erasing itself only leaves another trace is an involuted contact zone between species (penguin and human); it is a frontier beyond which lies nothing. Ponting photographed on ice at a crossroads in humanity's self-perception. The image evokes a multitude of questions: Is it a retrospect or prospect? Accident or design? Are we (or they) coming or going? Is this "X" the map—or the territory? Maps impose recursive meaning: The map refers to the territory and the territory owes its intelligibility to the map. The tracks of the sled like the violence of the visual impose order from above and from the perspective of coming after. Together, they produce ice as a target. X marks the spot. This target, as the post-structuralists would say is also under erasure. In a fit of artistic petulance, Ponting makes a grand, illiterate signature, a furious X signifying nothing. At the hinged crux of lines of sequential and overlapping expeditions both preceding his and stemming from his, Ponting had come all the way with his compass, ink, and lenses—only to discover ice.

But in the materialist frame that *Antarctica as Cultural Critique* has worked to create, nothing can be erased. Ice time is not reversible; it is not recursive, leading always back to its own origin. Memorialization, the looking at old photos, and yes, even the readings that have been produced so self-consciously by this book cannot unentrain the processes of ice. Ice cannot be undone.

Ponting's abject mystery on ice has one last entanglement with the academic frame of this book; it points to the field of American Studies as it is no longer founded on a notion of America that is a consequence of an overly secured, too-knowable past.

Notes

Introduction: *On Ice*

1. This post-structuralist insight comes from Jean Ricardou, "The Peculiar Character of the Water." *Poe Studies* IX(1) (June 1976): 1–9. Holes at the Poles or Hollow Earth theorizing of John Cleves Symmes has its critical center in early nineteenth-century America and as a naturalistic overlap with Antarctic fantasies. See chapter 2, for example, of David Standish, *Hollow Earth: The Long and Curious History of Imagining Strange Lands, Fantastical Creatures, Advanced Civilizations, and Marvelous Machines Below the Earth's Surface.* Cambridge, MA: Da Capo Press, 2006.
2. Antarctic Studies in the twentieth century has been dominated by science disciplines. Recent years have seen the growth of two major interdisciplinary degree-granting programs with Canterbury University in New Zealand's Gateway Antarctica and the Scott Polar Research Institute at Cambridge University in the United Kingdom. Cultural studies and visual culture approaches to the Polar regions are evidenced by occasional conferences and collections, including "Imagining Antarctica" in 2008 held at Canterbury University, New Zealand, and "Antarctic Visions" in 2010 in Hobart, Tasmania.
3. Robin Mckie reports in *The Guardian* June 9, 2012, that Douglas Russell, a curator of birds at the British Natural History Museum has unearthed a previously suppressed report on the behavior of Adelie penguins by George Murray Levick, a scientist on Robert Scott's 1910–13 Antarctic expedition. Apparently, Levick had been so distressed by observing behaviors of the male penguins that seemed to him as rape, necrophilia, gang rape, child rape, and murder that he wrote up his field notes in Greek so no innocent reader should be affronted by the truth about penguins. But now we know.
4. The novelist Tom Wolfe in a January 30, 2005, op-ed piece in *The New York Times*, "The Doctrine That Never Died," on the uncanny afterlife of the Monroe Doctrine, refers to the "curious case" of Antarctica, a non-claimed territory that might be south of South America, in the course of critiquing the Bush administration's deployment of exceptional power in the build up to the Gulf War.

Wolfe's linking of US Antarctic metageographies (if not policies) to US empire is precisely what American and Antarctic exceptionalisms seek to suppress.

5. Two books by Timothy Morton, *Ecology without Nature: Rethinking Environmental Aesthetics* (2007) and *The Ecological Thought* (2010) develop "dark ecology," Morton's term for an ecological mesh that precludes conventional nature/culture dualities that found environmentalism.

6. Peter Krapp (2012) describes the technoscientific diminishment of a science-run Antarctica, as a "beaker-utopia" that has replaced the mysteries of oral tradition with "measurement, and extrapolations, prognoses and doomsday scenarios" (151).

7. Lisa Bloom's 1993 *Gender on Ice* has been the most thorough and influential exposure of the gendered and raced power dynamics of polar exploration. Routinely cited in discussions of multiculturalist reassessment of polar exploration, less understood is its final section, "Setting Things Alright: Technology, the Gulf War, and Peary." Anticipating much of the criticism of the empire of turn-of-the twenty-first-century American Studies, Bloom argues that the masculine melancholy of the polar explorer was reinvigorated during the US Persian Gulf War of 1991. Linking technologies of gender, race, and science writing, Bloom reveals the endless serviceability of masculine failure (in polar exploration) as an alibi for the now permanent forms of loss and disorganization instrumentalized if not imposed under US neoimperialism.

1 Antarctic Convergence: The Problem of Antarctic Mapping

1. See Klaus Dodds, "Antarctica and the Modern Geographical Imagination (1918–1960)."

2. John Wylie, "Becoming Icy," and Kathryn Yusoff, "Antarctic Exposures," both consider the production of Antarctic territory in terms of the failing male body.

3. The term landscape, of course, references cultural constructions; in fact, no one has ever been able to gaze in the geographical direction of the Antarctic without some sense or other formed by specific aesthetic regimes. This is made clear by Paul Shepheard's inclusion of Antarctica under the chapter heading "Hope" in his *The Cultivated Wilderness.* Furthermore, those ways of seeing are inevitably attached to nation. My point here is simply that Antarctica's relative marginality in human history has resulted in a humanistic "gaze" inadequate to the territory. As far as accounts of Antarctica as part of history as normatively understood, there are a number of readable histories of Antarctic exploration and discovery such as Chapman, *The Loneliest*

Continent as well as a wealth of monographs and journals on the scientific, legal, and leisure associations attached to the region. Sahurie in *The International Law of Antarctica* and Beck in *The International Politics of Antarctica* include helpful bibliographies. Antarctica, a "travel survival kit," provides a nonspecialist yet up-to-date and nicely illustrated information on a range of Antarctic topics.

4. "Epistemological security" implies various Enlightenment legacies: the economic formation of capitalism in the West, the epistemological ascendancy of humanism, and the ideological rise of individualism and representative democracy. Of course, even these as references to "modernity" have been under critique in contemporary critical theory. See, for instance, Michel Foucault, *Discipline and Punish; The Order of Things;* and "What is Enlightenment?"; as well as Jurgen Habermas, "Modernity—An Incomplete Project." For more general conversation about the Enlightenment, see two books by G. S. Rousseau: *The Languages of Psyche: Mind and Body in Enlightenment Thought* and *Enlightenment Crossings: Pre- and Post-modern Discourses, Anthropological.*

5. For an important volume on the relation among vision, technology, modernism, and modernity, see Hal Foster, ed., *Vision and Visuality.*

6. Some of the most significant critical commentary on the "virtual" includes: Allucquere Rosanne Stone, *The War of Desire and Technology at the Close of the Mechanical Age;* Claudia Springer, *Electronic Eros: Bodies and Desire in the Post-Industrial Age,* and Donna Haraway, "A Manifesto for Cyborgs: Science, Technology, and Socialist Feminism in the 1980s."

7. The ATS has since its beginning been dominated by the industrialized and postimperial nations. In 1982, the unlikely nation of Malaysia, which through geography and history has never managed a traditional legal territorial claim based on discovery, conquest, or settlement, or even one based simply on nearness, protested this state of affairs. Malaysia argued that the neocolonialist system in which former imperial powers continue to profit at the expense of their former colonies under the guise of global capitalism must cease, and that one gesture toward a more equitable future among nations would be the inclusion of former "resource" nations as signatories to the ATS. As of the present, Malaysia and other newly industrializing nations still have no claim on the resources held in deferral by the ATS. For a fuller discussion, see Suter and Klotz. On the ATS, see Mhyre, *The Antarctic Treaty System: Politics, Law, Diplomacy;* Peterson, *Managing the Frozen South: The Creation and Evolution of the Antarctic Treaty System;* Sahurie, *The* International Law of Antarctica; and Jorgensen-Dahl.

2   Refusing History after Ursula K. Le Guin's "Sur"

1. The most important expressions of this linkage of failure and heroism in Anglo-European polar exploration are found in Bloom (1993)
   and Spufford (1996).
2. Although the Scott expeditions were bad exploration, they make a
   good story. Books and articles on Scott overwhelmingly outnumber those on Amundsen. In the realm of fiction, the attention continues to go to Scott's tragic expedition. The figure of Scott has
   become linked in modern south polar imaginings from the "white
   road" passage in T. S. Eliot, *The Wasteland* (1929) to Crispin
   Kitto, *Antarctica Cookbook* (New York, : St. Martins Press, 1984),
   in which a man escapes into fantasies of living in Scott's hut on
   the Antarctic cape. Almost all the films and videos on the history
   of Antarctica devote disproportionate attention to Scott's expeditions. With such recent and popular offerings as Beryl Bainbridge's
   The Birthday Boys (New YorkLondon: Gerald Duckworth, 1991),
   an imaginative retelling of Scott's last expedition, it is clear that
   Anglo-American interest in the saga has not yet waned. For a discussion of British mythologizing of Scott, see Peter J. Beck, "The
   Legend of Captain Scott 75 Years After," *Polar Review* (Summer
   1985): 604–619.
3. Le Guin writes in the tradition of Antarctic hoax in which an anonymous author/explorer presents a tale of a marvelous discovery which
   (1) cannot by known science be either proved or disproved, or (2)
   presents an alternate view of, or unknown, history of the event, most
   often through the discovery of a "lost" manuscript. Some examples
   are Poe, *The Narrative of Arthur Gordon Pym*; Charles Romyn Dake,
   A Strange Discovery (1899), and H. P. Lovecraft, *At the Mountains of
   Madness* (1936).
4. List pro- and anti-Scott books. Is there a list somewhere? Also, web-
   sites and ongoing chat.
5. On Antarctica as rupture to modernism, see Glasberg (1997),
   Chapter 1, and Ziskind (2007).
6. Ever since Roald Amundsen reached the South Pole in December
   1911, followed closely by Scott of Britain, the race to stake out
   geographical claims in the Antarctic has remained steady. Edmund
   Hillary and Vivian Fuchs used mechanical means to cross the continent in 1958, while Reinhold Messner was first to ski solo across
   Antarctica in 1990. It is within this context that this most recent
   crossing must be placed.

   American Ann Bancroft and Norwegian Liv Arnesen became
   the first women to cross the Antarctic continent in a mechanically
   unaided trek. They began their attempt in October 2000, reached
   the South Pole on January 16, 2001, and ended their adventure on
   February 18, 2001. Bancroft and Arnesen, veterans of previous polar

expeditions, had funding from corporations such as Volvo, Motorola, and Pfizer and returned to much major media attention, appearing on Good Morning America and NBC's Nightly News, to name just a few. Every aspect of their journey had been tracked by a sophisticated website (www.yourexpedition.com). Their expedition actually came short of its goals (they had to abandon efforts 500 miles short of their original goal), even while two Norwegian men (Erik Sonneland and Rolf Bae) trekking at the same time as the women's team, made the entire continental traverse, but without the elaborate communication organization. For more information on the adventure aspects of the crossing and comparisons to other crossing endeavors, see John Howard, "The Southern Traverse" in *Adventure* (May/June 2001): 31–32.

7. I do not want to overemphasize the analogy of feminism and subaltern studies in any way, not even in terms of the more specific topic I take up here, of temporality. Feminist and subaltern objects of study (the subjectivities that serve as objects of the discourses), and the very discourses in turn created by the emergence of the academic fields, are distinct; I am suggesting neither a shared nor an analogous historical development. I am, however, interested in certain intersections and self-conscious inter-awarenesses between the two discourses.

8. See Mary Louise Pratt, *Imperial Eyes: Travel Writing and Transculturation* (London and New York: Routledge, 1992) on the "continental turn" in geography in conjunction with the modernist aspiration to totalization, or the move toward a rational, interconnected, and hierarchicalized world knowledge. For the idea of modernity and space and time, see Stephen Kern (1983). For the idea of the social mapping or creation of continents within the mapped globe, see Lewis and Wigen (1997).

9. See Susan Stanford Friedman, *Mappings: Feminism and the Cultural Geographies of Encounter* (Princeton, NJ: Princeton University Press), 1998, p. 207.

10. Susan Gubar (1981: 243–263) offers a detailed and persuasive understanding of the long history of the imagery linking blankness and female endeavor/creativity/corporeality. Initially noting that the male artist has tried to usurp female potency through myths of male creation so to "evade" the "humiliation...of acknowledging that it is he who is really created out of and from the female body" (243), Gubar goes on to describe the difficulty the woman artist (or in this case, the explorer) has in trying to join a tradition of specifically male creativity and endeavor that equates pen with penis and insists that women be either passive receptacles of male genius, muses to male creativity, or works of art in themselves—the central metaphor of the "blank page."

11. This technique of sighting and tracing man-made marks in the ice, essentially an image of following oneself because there are no other

recognizable reference points, is matched by the name of a navigating device, the "artificial horizon" as an image of the projection of the values and systems of understanding of the explorer's culture onto an empty land.

12. Le Guin is referencing artifacts of exploration history with which her readers are likely to be familiar. Both the Scott and Shackleton huts have been preserved and refurbished for national pride and the Antarctic tourist trade. The huts are treated almost like shrines to which visitors make a pilgrimage. Photographs of the huts, their contents, and even of the men of the original Scott and Shackleton expeditions have been widely available, since all early expeditions took along their own photographers. See especially Herbert Ponting, *The Great White South* (London: Gerald Duckworth and Co. [1921], 1999).

13. The ozone hole is arguably among the most well-known features of the Antarctic in the popular imagination. Evidence of a hole opening in the ozone above the south polar region was first brought by a team of atmospheric scientists led by Dr Susan Solomon, the work for which she was recognized with the US National Medal of Science. Solomon has moved from her association with the ozone hole phenomenon to entering the Scott–Amundsen debates. Using her scientific training in atmospherics, Solomon has recently published a defense of Scott's methods and leadership skills, arguing that his failure was due to unusual weather patterns and not to bungling, as many others had maintained. See Solomon (2001). Although the latest scientific evidence shows that the ozone hole is closing somewhat, the crisis of the environment in Antarctica remains one of the defining features of the region and one that plays a central role in present and future governance.

14. Of course, all geographical markers such as longitude and latitude, the equator, and national borders and borders of all kinds are equally artificial. What is interesting about the Antarctic as a setting is its placement as geographically and historically last, a coming at the end of the world and the end of time. The South Pole as the convergence of time and space is not any more artificial than other points on the mapped earth, but it does hold a heightened significance within a metageography of empire, a symbolic valence exploited by Le Guin in "Sur."

15. See Jehlen (1981: 575–601) for an early and important articulation of the problem of critiquing history from a position necessarily within history.

    The debate, roughly drawn, between post-structuralist and traditional scholars continues within the disciplines. For a good discussion of the "post-structuralist turn," see Michelle Barrett and Ann Phillips, eds, *Destabilizing Theory: Contemporary Feminist Debate* (Stanford, CA: Stanford University Press, 1992). For an example of feminist debate over such terms as evidence, experience, and the body, see Kathleen Canning, "History after the Linguistic Turn: Historicizing Discourse and Experience," *Signs* 19(2) (1994): 368–404, for a discussion of the debates set off by Joan Scott in "Gender: A Useful Category of Historical Analysis," *American Historical Review* 91.5 (1986): 1053–1075 and *Gender and the Politics of History* (New York: Columbia University Press, 1988). Also, Joan Scott, *Feminism and History* (Oxford and New York: Oxford University Press, 1996).

16. Feminism, like other discourses, has had to become aware of the problem of its historical complicity with first-world imperialism. Specifically, "[i]n supporting the agendas of modernity, therefore, feminists misrecognize and fail to resist Western hegemonies" (Grewal and Kaplan 1994: 2–3).

17. Le Guin's tale also suggests another counterclaim to history by elevating "fable" to coeval status with historical narrative even while insisting on the non-corroborationality of fable, including the enveloping narrative/fable "Sur." When the members of the women's expedition redescribe their exploits as "fairy tales" only to their children and grandchildren, Le Guin troubles the distinction between origin history and origin fables. History in this case is the product of European culture while fable occupies a shadowy, less-distinguished status associated with colonized cultures.

18. The major source on Shackleton as hero and leader of men is Huntford (1986). The story of Shackleton's leading his men to survive for two years in the sub-Antarctic regions has been taken up at several levels of popular culture. The New York Museum of Natural History mounted an exhibition of Shackletonia in 1999 and George Butler directed two 2001 documentaries: *The Endurance: Shackleton's Legendary Antarctic Expedition* and a shorter IMAX version, *Shackleton's Antarctic Adventure*. Both films are unabashed hagiographies of Shackleton the man and leader of men. Rumor is that a Hollywood feature film on the expedition and rescue is underway. A business-oriented publication, *Shackleton's Way* (Margaret Morrell and Stephanie Capparell; London: Nicholas-Brealey, 2001), proposing Shackleton as an exemplary manager of men from whom contemporary business people have much to learn all attest that the present is a moment of nostalgia and re-remembering of this disaster. Focus on the managerial and human success aspect of Shackleton's disaster allows Anglophiles for whom Scott's loss to Amundsen in the race to the pole has never sat right to promote another British explorer—who did not achieve the pole, but at least did not die trying.

19. Chile has pursued an aggressive counter-European agenda for Antarctic territory that includes a rich cultural history and legal claims. See Dodds (1997) and Howkins (2006).

3   "Who Goes There?": Science, Fiction, and
US National Belonging in Antarctica

1. Quoted in Walter Chapman, *The Loneliest Continent: The Story of Antarctic Discovery* (Greenwich 1964), p. 31. Countless writers on Antarctica begin their narratives of Antarctica's (European) exploration citing Cook's premature sense of an ending to earth's southern geography.

2. Books on Antarctica typically open with a gesture to the continent's earliest speculative conceptualizations by the Egyptian Ptolemy, whose second-century map of the world introduced the area he labeled *terra australis incognita*. See, for example, P. I. Mitterling, *America in the Antarctic to 1840* (Urbana, IL: University of Illinois Press, 1959), p. 4.

3. See Stephen Pyne, "The Extraterrestrial Earth: Antarctica as Analogue for Space Exploration," *Space Policy* 23 (2007): 147–149, for a discussion of the limitations of human inhabitation in Antarctica and its repercussions for science policy and cultural development, including resource extraction, in Antarctica. Pyne's exceptionalism is also discussed in Chapter 2.

4. The desultory history of US government–sponsored expeditions throughout the nineteenth and twentieth centuries attests to the way that Antarctica's possibilities only appealed during periods of relative national inactivity. After the Wilkes Expedition team returned in 1842 with views of the icy reality of the southern oceans and the Civil War era arose, it was not until the turn of the twentieth century that another US pioneering oceanographer Matthew Maury's oceanic endeavors took place. In the twentieth century, the greatest US Antarctic explorer, Byrd, filled in the gap between the world wars with his exploits. By the time the United States asserted its global mastery after WWII, the nation's goals for Antarctica consolidated around science as a means of détente and controlling particularly Soviet interests.

5. For an assessment of Antarctic tourism, see Mason, *The Growth of Tourism* (2000): 358. For a more complex argument on the interrelation of tourism, self-regulation of the industry, and activity on the continent more broadly, see Murray and Jabour, 309–317. On the ATS, see Stokke and Vidas (1996).

6. Competition and strategizing over resources and territorial claims in Antarctica has recently reemerged as an almost daily international news topic. See for example, "Nations Chase Rights to Lucrative Antarctic Resources," *The Epoch Times* http://en.epochtimes.com/tools/printer.asp?id=65009. Accessed 4/16/08.

7. Throughout the 1990s, routine praise for the Treaty's ability to negotiate the needs of the most powerful signatories and to control the challenges of upstarts has been somewhat balanced by scholarship discussing persisting problems, including the Treaty's lack of ability to regulate tourism. See for example, A. Jorgensen-Dahl and

W. Ostreng (eds), *The Antarctic Treaty System in World Politics* Oslo: Fridtjof Nansen Institute, 1991), and Beck (2004): 205–212, for tensions developing between original signatory states and emerging states and global interest, and Dodds (2006): 59–70 for a geopolitical assessment that names the condition of the postcolonial. More recent accounts have been more critical of the ATS, not as a treaty in itself, but as it has materially been able to control or even predict human intervention and change in the region. See for example, Scott (2003): 473.

8. Driver (2001). "Unashamed heroism" (4) is the fantasy of a consequence-less and endless imperial process that by the time of Antarctica's technical availability to humanity was being self-consciously lamented.

9. British versions of the Scott legend oscillate from the hagiographic *Scott of the Antarctic*, Carl Frend, dir., 1949, to Monty Python's parody *Scott of the Sahara*, Flying Circus TV Show, episode 23, 1970, to Beryl Bainbridge's purposefully disjointed yet sympathetic *The Birthday Boys*, London, 1991.

10. For the neo-foundational reassessments of the role of repressed empire in the American imaginary and of the transnational discussion, see Donald Pease and Amy Kaplan (eds), *Cultures of United States Imperialism* Duke University Press: Durham, NC, 1993.

11. For recent discussions of internal and external rearrangements of American Studies within a global field imaginary, see Paul Giles, "Commentary: Hemispheric Partiality," *American Literary History* 18(3) (2006): 648–665, and B. T. Edwards, "Preposterous Encounters: Interrupting American Studies with the (Post)Colonial, or *Casablanca* in the American Century," *Comparative Studies of South Asia, Africa, and the Middle East* (2003): 70–86. On how ideologies have produced a particular rendering of the world map or "metageographies," see R. Lewis and K. Wigen, *The Myth of Continents: A Critique of Metageography*. Berkeley: University of California Press, 1997).

12. On October 17, 2007, Britain officially pressed its geographic claims to three areas of Antarctica: www.guardian.co.uk/news/2007/oct/17/antarctica.sciencenews

13. While the United States operates the largest station, McMurdo, and arguably the most strategically placed one at the South Pole, the number of US base personnel (477 at McMurdo) is dwarfed by the Argentine and Chilean personnel (417 and 224, respectively). Although the US economic investment in Antarctica outpaces that of all other individual nations, economic investment as an index of cultural impact of an individual nation is not reflected in these numbers. Figures gathered from the Central Intelligence Agency factbook: www.cia.gov/library/publications/the-world-factbook/geos/ay.html#People

14. See for example, Diane Belanger (2006). Stephen Pyne (1986) contains an excellent account of US-centered geopolitics from Byrd's 1928 flight over the South Pole. Pyne argues for Byrd's prescient internationalism and global vision focused on Antarctica.

15. Variations on the theme of Antarctica as "icebox for the world" appear throughout Byrd's papers. See Ohio State University Archives. Papers of Admiral Richard E. Byrd, RG 56.1, folder # 3850, 3523. Byrd never entirely escaped his own or his constituents' desires that Antarctica might fulfill US desires for new land and investment opportunities. Ironically, a posthumous article (actually a republishing of a piece Byrd produced in 1957, toward the end of his life) in *Over Here: A Monthly Journal of Afterlife Consciousness* 1V(3) (1960) revives the frustrating search for a useful Antarctica in its title: "Antarctica is Lush with Resources Says Rear-Admiral Byrd" through the discourse of para-scientific spiritualism.

16. Despite both speculation and hindsight, it seems now clear that the United States never developed a unified vision of its Antarctic territory before the International Physical Year (IPY) of 1958. For a fact-filled yet strangely nationalistic lament for US Antarctic policy, see Moore (2004): 19–30. The Byrd archive at Ohio State University offers an admittedly selected yet illuminating glimpse into public support for a national territorial claim in Antarctica. Byrd's exploration exploits inspired poetry linking him to Columbus, and overall, the letters from the public indicate a desire on the part of citizens for the explorations to translate into possessions that echoed national founding myths.

17. Nor could any other nation's acts qualify, including the Third Reich's New Schwabenland claim anchored by swastika droppings. M. Chabon's *The Amazing Adventures of Kavalier and Clay* New York: Picador, 2000) invents an Antarctic US military installation (modeled on actual US–German tension in Greenland) and an encounter with a rival German fighter: WWII is thus played out in Antarctica.

18. Paul Carter (1979) titles his chapter on Advance Base "To Walden Pond with Gasoline Engines," alluding to Thoreau's experiment in living on the American land and cites evidence for how Byrd was placed in a line with representative American men including Jefferson, Edison, and Twain in school anthologies, p. 181.

19. R. E. Byrd, *Alone*, Afterword by D. G. Campbell. New York: Kodansha International, 1995).

20. Stephen Pyne, *The Ice*, p. 190. Pyne notes Byrd's oscillation between science as justification for Antarctic presence and a more spiritual benefit for humanity, but he does not elaborate.

21. Ohio State University Archives. Papers of Admiral Richard E. Byrd, RG 56.1, folder # 2756.

22. Of course, within the hegemony of science, many other forces contend. While the NSF is the de facto authority for US activities

associated with its science agenda (which accounts for the majority of human activity on the continent), the NSF itself operates the Antarctic Artists and Writer's Program and a journalist program in order to foster non-scientifically trained persons' knowledge and reporting. The majority of the people working in Antarctica are in fact nonscientists. The substantial tourism industry as well fosters nonscientific routines and action. For a more critical assessment of US science in Antarctica, see J. Spillers, "Re-imagining United States Antarctic Research as a Defining Endeavor of a Deserving World Leader: 1957–1991," *Public Understanding of Science* 13 (2004): 31–53.

23. There is much work to be done in approaching this question of the redeployment of science as a territorial, spatial force, and some of it has been underway, for example, Christy Collis and Quentin Stevens, "Cold Colonies: Antarctic Spatialities at Mawson and McMurdo Stations," *Cultural Geographies* 14 (2007): 234–254. For an account of the construction through the performance of the law of Australian Antarctic space, see C. Collis, "The Proclamation Island Moment: Making Antarctica Australian, *Law Text Culture* 8 (2004): 1–18.

24. D. A. Stuart, *Astounding Science Fiction* (1938): 62.

25. One exception is Elle Leane, "Locating the Thing: The Antarctic as Alien Space in John W. Campbell's 'Who Goes There?'," *Science Fiction Studies* 32(2) (2005): 225–239. Arguing that the Thing "serves as the embodiment of the continent itself," Leane understands Antarctic space not as geographic but as a body under abjection, which is precisely the point of A. M. Butler's reading of the 1982 John Carpenter film remake *The Thing* in "Abjection and *The Thing*," *Vector* 24(3) (2000): 10–13.

26. Jason Kendall Moore, "Bungled publicity: Little America, Big America, and the Rationale for Non-claimancy, 1946–61," *Polar Record* 40 (2004): 19–30, cites a 1946 proposal to nuke the Antarctic ice cap to get at its minerals and the subsequent *New York Times* op-ed piece pointing out the disaster of a rise in sea levels (p. 21). The discussion of the possibilities of anthropogenic change to the ice cap and its effects is eerie given the reality of climate change today.

27. *The Blob* (1958 dir. Yeaworth) resolves with the Arctic used as a frozen waste container for the alien threat. The final frame of the film is a question mark superimposed over the Arctic tundra as a helicopter sling-load dumps the blob out on to the ice.

28. See P. Beck above and K. Dodds, *Geopolitics in Antarctica: A View from the Southern Rim* (London, 1998).

29. The film has become a cult favorite and enjoys an elevated reputation at South Pole Station, where workers consider it the only film to capture the *feel* of living in Antarctica. In 2005, I gave a presentation on *The Thing* in the galley at McMurdo Station in Antarctica

to an albeit captive audience. People were not much impressed by what an academic had to say, but we had a lively conversation about the accuracy of the depiction and the effect of obvious embellishments or retentions from the 1938 military base version, such as the characters' possessing firearms (much less flamethrowers), which have been long banned from the demilitarized continent. The long-awaited remake finally arrived in theaters in 2011. See Glasberg 2012.

30. For a social science assessments of *The Thing* as remake, see M. Katovich and P. Kincaid, "The Stories in Science Fiction and Social Science: Reading *The Thing* and Other Remakes from Two Eras," *Sociological Quarterly* (1993): 619–639. For an excellent filmic analysis of medium and remakes, see P. Crogan, "Things Analog and Digital," *Film and Philosophy* (2001): 13–23, as well as S. Kneale, "You've Got to Be Fucking Kidding!': Knowledge, Belief, and Judgement in Science Fiction," in: A. Kuhn (ed.), *Alien Zone: Cultural Theory and Contemporary Science Fiction Cinema* (New York and London, 1990).

31. Neoimperialism goes by other names such as "informal empire," "colonialism without colonies," or "cultural imperialism." While all these terms imply subtle distinctions, they all bear the burden of describing new modes of consolidating power after the end of the "territorial imperative" that marked sixteenth- to nineteenth-century imperialism. With so little new territory available to occupy and with new economic tools and international agreements keeping outright wars over territory mostly in check, expansion of power and capital has followed nonterritorial paths. New ways of expanding capital and amassing power exploit debt, the operations of transnational corporations, powerful international media and markets, and even environmental concerns to benefit the more powerful states. How a colony is defined under neoimperialism has expanded to include nonterritorial or deterritorialized peoples, places that are not strictly territories such as outer space, the deep sea, and Antarctica. Antarctica as the only unclaimed territory left on earth poses a challenge to such definitions. The term neoimperialism therefore is ironic and yet productive.

32. Recent assessments of the tangled histories of science and politics include S. Naylor et al., "Science, Geopolitics and the Governance of Antarctica," *Nature* (March 2008): 143–145 and F. Kormo, "The Genesis of the International Geophysical Year," *Physics Today* (July 2007): 38–43.

33. Information on Raytheon's Polar Program comes from its webpage at: www.raytheon.com. My arguing for the significance of a corporate military contractor subtending contemporary US Antarctic presence does not rely on a particular account—or critique—of Raytheon's practices. Rather, I am noting a historical shift in the

configuration of a US Antarctic from outside the borders of compliance within the ATS.

34. Chalmers Johnson, "America's Empire of Bases," Jan 2004.http://www.globalpolicy.org/empire/intervention/2004/01bases.htm.

35. While this chapter suggests a reading of US empire in the interstices of representation, policy, and physical presence, an excellent analysis of US cultural imperialism in films is by John Hegglund, "Empire's Second Take: Projecting America in *Stanley and Livingstone*," in: H. Michie and R. R. Thomas (eds), *Nineteenth Century Geographies: Anglo-American Tactics of Space* (New Brunswick, 2002), pp. 265–277, in particular his discussion of how representation enacts a new form of imperialism without a need for territorial acquisition: "the shift from map to film as the most culturally resonant representation of geographical space and, second, the transition from the British form of imperialism based on territorial acquisition to a United States form of imperialism based on the manipulation of image and spectacle...[*Stanley and Livingstone* 1939] justifies a more mobile, influential, 'disinterested' global presence, paving the way for an empire that could prosper without imperialism" (267).

36. Aant Elzinga, "Antarctica: The Construction of a Continent by and for Science," in: Crawford and Sinn (eds), *Denationalizing Science: The Contexts of International Scientific Practice* (Kluwer, 1993) argues that one by-product of international science is the "partial foreclosing of alternative concepts and approaches" (98). A self-published report dated August 17, 1988 by Bruce Manheim, Jr. of the Environmental Defense Fund (Washington, D.C.), *On Thin Ice: The Failure of the National Science Foundation to Protect Antarctica*, is a reminder of relatively early criticism of the science regime of the ATS in terms of its environmental negative effects.

37. Nicholas Johnson, *Big Dead Place: Inside the Strange and Menacing World of Antarctica* Port Townsend, WA: Feral House, 2006), is the only extended treatment of worker culture in the contemporary US McMurdo base, discussing Raytheon's personnel policies and the way they clash with non-US territorial service and with US-based tax law, and individual worker rights.

38. Many versions of the McMurdo peace sign exist. An exemplary one, named the "human peace sign," remains available at the progressive political website "Common Dreams": www.commondreams.org/headlines03/0119-02.htm.

39. The legal status of US citizens working in Antarctica has been complicated by unclarity about how to understand the US bases in Antarctica. Are they US territory or foreign? The answer has serious tax consequences for workers. Their struggles, eventually denied in 2010, to declare Antarctica foreign territory and thus to be exempt from federal taxation can be followed online: http://signon.org/sign/fair-taxation-for-contract.fb1?source=c.fb&r_by=1541123.

4 On the Road with Chrysler: Virtual Capitalism
and Empire without Territory

1. Information about the road to pole is available—I am embarrassed to report—on *Wikipedia*: http://en.wikipedia.org/wiki/South_Pole_Traverse. The reporting on the project has been limited to publications suported by the corporations that are involved in building the project. The *Wikipedia* site provides basic factual information. Official rationale for the road is to save on fuel, the greatest expense in science support, as well as to remain within self-regulated guidelines on environmental impacts.

2. Unfortunately, I was not able to secure permissions for this advertisement image, or for the other images discussed in this chapter. The use of Antarctica or its associated features, penguins and icebergs, is quite common and the images themselves are ephemeral. Collecting or surveying these types of images is not at all the intention of this chapter or the book as a whole, but quite the opposite. In discussing texts, whether literary, commercial, or simply popular, I am questioning the creation of a field or a history of Antarctica. Nevertheless, having been able to print the images would have made this chapter in particular more enjoyable to read.

3. The phrase is Robert Smithson's. See *Collected Writings of Robert Smithson*.

4. Fredrick Jameson in *Archaeologies of the Future: The Desire Called Utopia and Other Science Fictions*. London and New York: Verso, 2005, puts the problem of futurity in terms of industrial capitalism "our imaginations are hostages to our mode of production" (5). Amy Kaplan in "Imperial Melancholy in America." *Raritan* 28(3) (Winter 2009): 12–20, places US empire as a perhaps inevitable repetition of earlier empires falling to ruin.

5. For a provocative and trenchant expression of concern about the power of science management in Antarctica, see "millions go cold and hungry because of polar scientists" by Simon Jenkins in *The Guardian*. March 14, 2008.

6. A tendency among scholars working on the polar regions is to analogize the Antarctic situation based on the Arctic. The Arctic is oceanic, has many natives, and most importantly is in the large subject to territorial possession and thus future claims. For a précis of current issues, see Bloom, Glasberg, and Kay "Introduction: Gender on Ice" (2008).

7. See Glasberg 2000 for a more sustained discussion of the Chrysler advertisement in the context of transnational American Studies.

8. That Chrysler Corporation required another US government bailout after the 2008 global financial collapse is a perfect example of the repetitve structures of capitalism's drawn-out ransacking of taxpayer money. And in terms of this chapter's focus on representational strategies in advertising and corporate involvement in international science, it is beyond the scope.

5 Photography on Ice

1. Though Porter's Antarctic landscapes preceded the 1993 anti-mining agreement under the ATS, pressure to extract value from increasingly scarce territorial resource has only increased since 1978.

2. See Bright (1991) for a description of the Sierra Club aesthetic. For a fuller background on Porter's nature ideology, see Dunaway (2004); also see the Special Issue of *Aperture* "Beyond Wilderness" (2005) for a "critique of representational practices of major conservation organizations such as The Sierra Club...and their...'pinup' photography."

3. The widely reproduced Scott photographs and the much-less-well-known Amundsen photographs depict haggard men in thick gear amid a scene of almost total whiteout. Their usefulness as evidence has been obviated by their ancillary nature. Collis (2002) discusses Mawson's use of photography as part of a claiming repertoire. Cameras came to Antarctic waters in 1895. Scott had cameras along on the 1905 expedition, as did Shackleton on his 1907 attempt to reach the pole.

4. Antarctica's challenge to human perception and the phenomenon of whiteout is discussed in Glasberg (2004), Fox (2006), and Noble (2008). For an art theoretical discussion of blankness, see the chapter on "Blankness as a Signifier" in Gilbert-Rolfe (2000).

5. See Amy Kaplan's discussion of American Exceptionalism in Kaplan and Pease (2000).

6. See Bloom, Glasberg, and Kay (2008) "Introduction: New Poles, Old Imperialism" for the way rising oil prices have increased pressure to exploit the polar regions as possible solutions.

7. Photographs of US explorer Richard E. Byrd's "Little America" bases from 1928 to 1938 document a built environment prominently featuring jumbled boxes and fuel cans on a field of ice. For a range of photographs depicting US bases throughout the 1930s and 1940s, see "All-out Assault on Antarctica," *National Geographic Magazine* CX(2) (August 1956): 141–180. Explicitly modeled after the Little America base, the opening shots of the Antarctic base in John Carpenter's 1982 *The Thing* pans across empty fuel drums against the ice.

8. A filmic/cultural icon of a "defunct social system" is the scene in *Planet of the Apes* (1972) depicting the Statue of Liberty as a beached ruin.

9. Pyne (1986) offers a portfolio of Antarctic representation including Ponting, Porter, and Emil Schulthess, p. 205.

10. The term refers to love of extremes as well as to a class of creatures that has evolved to thrive in desert conditions such as those prevailing in Antarctica.

11. In a scene from "Sur," Le Guin describes her characters encountering the area around Scott's abandoned hut. Looking much different

than the famous Ponting shots, the hut area is a mass of dog turds and open spilt containers, sealskins, and bloody guts. The narrator remarks: "The backside of Heroism is often sad; women and servants know that" (263).

12. In a sense remote sensing of the realist kind represented by Byrd's 1928 overflight of the South Pole had been prefigured in polar fantasy, the long history of mapping, as well as by the disembodied camera eye.

13. Jenkins, Simon. "Millions Go Cold and Hungry Because of Polar Scientists." *Guardian* (March 14, 2008): 14–15.

14. "After Nature" exhibit at New Museum, NYC 2008.

15. Cronin (1995), "The Trouble with Wilderness," in *Uncommon Ground*.

16. Earlier works include "Small Wars" (1999–2002), in which Lê photographed and participated in Vietnam War reenactments in South Carolina; and "29 Palms" (2003–04), in which United States Marines preparing for deployment play-act scenarios in a virtual Middle East in the California desert

17. Naomi Klein, *The Shock Doctrine: The Rise of Disaster Capitalism* (New York: Picador, 2007).

18. I too was a guest of the AAWP 2004–5.

19. For a discussion of professional versus amateur photography in Antarctica, see Glasberg (2007).

20. Although the sources for the images have not yet been identified, they are most likely from popular magazines of the late 1940s–1950s such as *Life* and *National Geographic*.

6 Sculpting in Ice: Climate Change and Affective Data

1. See Farley (2005) for an analysis on how these accounts are sometimes used by both the adventure travel industry and by business management consultants to promote men's leadership skills.

2. Reportage on Prince Harry's April 2011 North Pole charity trek was primarily in the fashion and entertainment area; for example: www.pinkisthenewblog.com/2011/03/prince-harry-does-british-gq-magazine. Spufford (1996) details the historical–cultural roots of polar exploration for the British.

3. For example, Barbara Hillary recently became the first African-American woman to reach both poles: www.womensviewsonnews.org/wvon/2011/02/79-year-old-african-american-woman-reaches-south-pole/

4. The 1959 Antarctic Treaty has "frozen" all historical territorial claims until 2042 in favor of a science-only policy open to wider international participation including postcolonial nations such as India and Malaysia and newer powers like China as well. On postcoloniality and Antarctica, see Klaus Dodds (2006); and, Christy Collis and Klaus Dodds (eds) (2008).

5. Sighting and following the tracks of previous expeditions is not only a standard technique in exploration, it is also a trope in written narratives as footsteps and footnotes occupy parallel planes of referentiality. That said, literalist versions of following-in-the-footsteps prevail. For a recent exemplary account, see "How to Retrace the 1912 Race to the South Pole," text by Joe Robinson; photographs by Geoff Somers: http://www.nationalgeographic.com/adventure/photography/adventure-dreams/south-pole-trek/lessons-learned.html.

6. For a representative collection of liberal feminist analysis of women in Antarctica, see Edwards (2003); Bloom (1993) argued that polar exploration in the late nineteenth and early twentieth centuries was integral to the social construction of a distinctive nexus of technological superiority and nationalism that was crucial to reifying a particular form of white masculinity. In the early twentieth century, both the North and South Poles represented one of the few remaining masculine testing grounds where "adventure and hardship could still be faced" (2). The role of women and people of color, in this vehicle for nation- and culture-building and the advance of scientific knowledge, was significantly elided at this historical moment. Also see Rosner (ed.) (2009).

7. Completed in 1975, the dome station eventually succumbed to the accumulation of ice and was replaced by a more conventionally designed, large, and modern station. The dome became an object of sentimentalized attachment and its dismantling by the NSF was resisted by many former pole workers. The dismantling was completed in January 2010. For nonofficial documentation, see: http://en.wikipedia.org/wiki/File:South_pole_dome_deconstruction.jpeg

Epilogue: Becoming Polar

1. Rarely reproduced, the image appeared in a 2005 catalog of the Maritime Museum accidentally reversed, with the dark landmass on the right, not left. Only by looking at Ponting's own print in his album held at the Scott Polar Research Institute did I realize the reverse.

Bibliography

"After Nature." Curated by Massimiliano Gioni. Exhibit at The New Museum, New York City, July–October 2008.

Agnew, John. *The Power of Place: Bringing Together Geographical and Sociological Imaginations.* Boston: UnwinHyman, 1989.

"All-out Assault on Antarctica." *National Geographic Magazine* CX(2) (August 1956): 141–180.

Alvarez, Al. "S & M at the Poles." *New York Review of Books* 54(14) (September 27, 2007):19–22.

Amy Kaplan. "Imperial Melancholy in America." *Raritan* 28(3) (Winter 2009): 12–20.

Arms, Myron. *Riddle of the Ice: A Scientific Adventure into the Arctic.* Norwell MA: Anchor Press, 1998.

Bainbridge, Beryl. *The Birthday Boys.* London: Gerald Duckworth, 1991.

Bancroft and Arnesen (www.yourexpedition.com).

Barrett, Michelle and Ann Phillips, eds. *Destabilizing Theory: Contemporary Feminist Debate.* Stanford, CA: Stanford University Press, 1992.

Barth, John. " 'Still Farther South:' Some Notes on Poe's *Pym*." In Richard Kopely, ed. *Poe's Pym: Critical Explorations.* Durham NC: Duke University Press, 1993, pp. 217–232.

Beck, Peter J. "Twenty Years On: The UN and the 'Question of Antarctica,' 1983–2003." *Polar Record* 40 (2004): 205–212.

———. "The Legend of Captain Scott 75 Years after." *Polar Review* (Summer 1985): 604–619.

———. *The International Politics of Antarctica.* Hampshire UK: Palgrave Macmillan, 1986.

Behrendt, John C. *Innocents on the Ice: A Memoir of Antarctic Exploration,* 1957. Boulder CO: University Press of Colorado, 1998.

Belanger, Dian. *Deep Freeze: The United States, the International Geophysical Year, and the Origins of Antarctica's Age of Science.* Boulder: University Press of Colorado, 2006.

"Beyond Wilderness." Special Issue of *Aperture* (2005).

"Blank Spots on the Map," *Time Magazine Review.* Available at: http://www.time.com/time/arts/article/0,8599,1876478,00.html

Bender, Barbara. "Landscapes are Experimental, Porous, Nested, and Open-ended." *Landscapes: Politics, Perspectives* (2002): 136–137.

Bloom, Lisa. *Gender on Ice*. Minneapolis: Minnesota University Press, 1993.

Bloom, Lisa, Elena Glasberg, and Laura Kay, "Introduction: New Poles, Old Imperialism?" *The Scholar and Feminist,"Gender On Ice Special Issue"* 7(1) (November 2008): http://www.barnard.edu/sfonline/ice/introduction_01.htm.

Bogues, Ronald. *Deleuze on Music, Painting, and the Arts*. London: Routledge, 2004.

Bright, Deborah. "The Machine in the Garden Revisited: American Environmentalism and Photographic Aesthetics." *Art Journal* 5(2) (1992): 60–71.

Brockmeier, Kevin. *The Brief History of the Dead*. New York: Vintage, 2007.

Butler, A. M. "Abjection and *The Thing*." *Vector* 24(3) (2000): 10–13.

Byrd, R. E. *Alone*, Afterword by D. G. Campbell. New York: Kodansha International, 1995.

———. "Antarctica Lush With Resources." *Over Here: A Monthly Journal of Afterlife Consciousness* IV(3) (1960): 1–4.

Byrd Archive at Ohio State—Byrd, Admiral Richard E. Byrd, RG 56.1, folder # 3850, 3523.

Campbell, David G. *The Crystal Desert: Summers in Antarctica*. New York: Houghton-Mifflin, 1992.

Campbell, John W. "Who Goes There?" (Pseudonyn Don A. Stuart). *Astounding Stories* (1938): 21–35.

Canning, Kathleen. "History after the Linguistic Turn: Historicizing Discourse and Experience." *Signs* 19(2) (1994): 368–404.

Carpenter, John, dir. *The Thing*. Paramount Pictures, 1982.

Carter, Paul. *Little America: Town at the End of the World*. New York: Columbia University Press, 1979.

Chabon, Michael. *The Amazing Adventures of Kavalier and Clay*. New York: Picador, 2000.

Chapman, Walker. *The Loneliest Continent*. Greenwich CT: New York Graphic Society, 1964.

Chaturvedi, Sanjay. *The Polar Regions: A Political Geography*. Chichester: John Wiley, 1996.

Chow, Rey. *The Age of the World Target: Self-Referentiality in War, Theory, and Comparative Work*. Durham: Duke University Press, 2008.

CIA factbook: www.cia.gov/library/publications/the-world-factbook/geos/ay.html#People

Collis, Christy "The Proclamation Island Moment: Making Antarctica Australian." *Law Text Culture* 8 (2004): 1–18.

Collis, Christy and Klaus Dodds. "Assault on the Unknown: The Historical and Political Geographies of the International Geophysical Year (1957–8)," *Journal of Historical Geography* 34 (2008): 1–14.

Collis, Christy and Quentin Stevens. "Cold Colonies: Antarctic Spatialities at Mawson and McMurdo Stations." *Cultural Geographies* 14 (2007): 234–254C.

Coole, Diana and Samantha Frost, eds. *The New Materialism: Ontology, Agency and Politics.* Durham NC: Duke University Press, 2010.

Cordes, Fauno. "A Bibliographic Tour of Antarctic Fiction." *AB Bookman Weekly* 82(21) (Nov 21, 1988): 2029–2036.

Cosgrove, Denis, ed. *Mappings.* London: Reaktion, 1999.

———. *Apollo's Eye: A Cartographic Genealogy of the Earth in the Western Imagination.* Baltimore: Johns Hopkins University Press, 2001.

Craciun, Adriana. "The Scramble for the Arctic." *Interventions: International Journal of Postcolonial Studies* 11(1) (March 2009): 103–114.

Crane, David. *Scott of the Antarctic: A Life of Courage and Tragedy.* New York: Knopf, 2006.

Crogan, P. "Things Analog and Digital." Special issue, *Film and Philosophy* (2001): 13–23.

Cronin, William. *Uncommon Ground.* New York: Harcourt Brace, 1995.

Dake, Charles Romyn. *A Strange Discovery.* New York: H. Ingalis Kimball, 1899.

De Mille, James. *A Strange Manuscript Found in a Copper Cylinder.* NY and London: Harper & Brothers Publishers, 1900.

Dean, Katrina et al. "Data in Antarctic Science and Politics." *Social Studies of Science* 38/4 (August 2008): 571–604.

Deleuze, Gilles. *Bergsonism.* New York: Urzone Books, 1988.

———. *Negotiations 1972–1990.* Trans. Martin Joughin. NY: Columbia University Press, 1995.

Dodds, Klaus. "Post-Colonial Antarctica: An Emerging Engagement." *Polar Record* 42 (2006): 59–70.

———. *Pink Ice: Britain and the South Atlantic Empire.* London and NY: I.B. Tauris & Co, 2002.

———. *Geopolitics in Antarctica: A View from the Southern Rim.* Chichester: Wiley and Sons, 1998.

———. "Antarctica and the Modern Geographical Imagination (1918–1960)." *Polar Record* 33(184) (1997): 47–62.

Dodds, Klaus and Christy Collis, eds. "Assault on the Antarctic." *Journal of Historical Geography: Special Feature on the International Geophysical Year* 34 (2008).

Dominic Sena, dir. *Whiteout.* Warner Brothers, 2009.

Driver, Felix. *Geography Militant: Cultures of Exploration and Empire.* Oxford: Blackwell Publishers, 2001.

Dunaway, Finis. "On the Subtle Spectacle of Falling Leaves." *Environmental History* (Oct 2004), Sciences Module 9.4.

Edwards, B. T. "Preposterous Encounters: Interrupting American Studies with the (Post)Colonial, or *Casablanca* in the American Century." *Comparative Studies of South Asia, Africa, and the Middle East* 23 (2003): 70–86.

Edwards, Kerry and Robyn Graham, eds. *Gender on Ice: Proceedings of a Conference on Women in Antarctica Held in Hobart, Tasmania, under Auspices of the Australian Antarctic Foundation.* Canberra: Australian Govt. Pub. Service, c1994. v, p. 129.

Eliot, T. S. *The Wasteland: Collected Poems, 1909–1962.* New York: Harcourt, Brace & World, 1963.

Elzinga, Aant. "Antarctica: The Construction of a Continent by and for Science." In *Denationalizing Science: The Contexts of International Scientific Practice,* Crawford and Sinn, eds. Dordrecht NL: Kluwer, 1993.

Eric Darnell, dir. *Madagascar.* Dreamworks, 2005.

ExecutiveBiz. "Interview with Bill Bodie." JD Kathuria on September 30.

Farley, Rebecca. "By Endurance We Conquer: Ernest Shackleton and Performances of White Male Hegemony." *International Journal of Cultural Studies* 8(2) (2005): 231–254.

Farman, Jason. "Mapping the Digital Empire: Google Earth and the Process of Postmodern Cartography." *New Media & Society* 12(6) (2010): 869–888.

Fend, Peter. "A Conversation with Peter Fend with David Joselit and Rachel Harrison." *October* 125 (Summer 2008): 117–136.

Foster, Hal. *Vision and Visuality.* New York: The New Press, 1988.

Foucault, Michel. *Discipline and Punish: The Birth of the Prison.* Trans. Alan Sheridan. NY: Vintage Books, 1995.

———. *The Order of Things: An Archaeology of the Human Sciences.* NY: Vintage Books, 1994.

———. "What is Enlightenment?" in *The Foucault Reader,* ed. Paul Rabinow. NY: Pantheon, 1984.

Fox, William. *Terra Antarctica: Looking into the Emptiest Continent.* San Antonio: Trinity University Press, 2005.

Fredrick Jameson. *Archaeologies of the Future: The Desire Called Utopia and Other Science Fictions.* London and New York: Verso, 2005.

Frend, Carl, dir. *Scott of the Antarctic.* British Studios, 1948.

Friedman, Thomas. *Hot, Flat, Crowded: Why We Need a Green Revolution and How It Can Save America.* New York: Farrar, Strauss, Giroux, 2008.

Friedman, Susan Stanford. *Mappings: Feminism and the Cultural Geographies of Encounter.* Princeton, NJ: Princeton University Press, 1998.

"Future Politics: An Interview with Kim Stanley Robinson." *Gender on Ice: Proceedings of a Conference on Women in Antarctica Held in Hobart, Tasmania, under Auspices of the Australian Antarctic Foundation,* 19–21 August 1993, compiled by Kerry Edwards and Robyn Graham.

George Miller, dir. *Happy Feet.* Warner Bros. Studio, 2006.

Gilbert-Rolfe, Jeremy. *Beauty and the Contemporary Sublime.* London: Allworth Press, 2000.

Giles, Paul. "Commentary: Hemispheric Partiality." *American Literary History* 18(3) (2006): 648–665.

Glasberg and Lisa Bloom. "Chapter 8: Disappearing Ice and Missing Data: Visual Culture of the Polar Regions and Global Warming." In *Far Fields: Digital Culture, Climate Change and the Poles*, Andrea Polli and Jane Marsching, eds. Chicago: Intellect Books, 2011.

Glasberg. "On the Road with Chrysler: From Nation to Virtual Colony." In *The Postcolonial U.S.*, Richard King, ed. Urbana: Illinois University Press, 2000, pp. 154–170.

———. "Camera Artists in Antarctica." *Photomedia: New Zealand Journal of Photography* No. 65 (2007): 21–23.

———. "Viral Things." In *Special Issue on Viral*, P. Clough and J. K. Puar, eds, *Women's Studies Quarterly* 40(1 and 2) (Spring/Summer 2012): 201–210.

———. "Bad Light: A Visual Approach to Antarctica." In *Breaking Ice: Re-Visioning Antarctica*. Wellington: Adam Art Gallery Te Pakata Toi Victoria University Press, 2005.

Grewal, Inderpal and Caren Kaplan, "Introduction: Transnational Feminist Practices and Questions of Postmodernity." In *Scattered Hegemonies: Postmodernity and Transnational Feminist Practices*. Minneapolis: University of Minnesota Press, 1994.

Grosz, Elizabeth. "Feminism, Materialism, and Freedom." In *The New Materialism*, Diana Coole, ed. Durham: Duke University Press, 2010.

Gubar, Susan. " 'The Blank Page' and the Issues of Female Creativity." *Critical Inquiry* 8(2) (Winter 1981): 243–263.

Habermas, Jurgen. "Modernity—An Incomplete Project." Trans. Seyla Benhabib. In *The Anti-Aesthetic: Essays on Postmodern Culture*, ed. Hal Foster. Seattle: Bay Press, 1983.

Haraway, Donna. *Simians, Cyborgs, and Women: The Reinvention of Nature*. NY and London: Routledge Press, 1991.

Hegglund, John. "Empire's Second Take: Projecting America in *Stanley and Livingstone*." In *Nineteenth Century Geographies: Anglo-American Tactics of Space*, H. Michie and R. R. Thomas, eds. New Brunswick: Rutgers University Press, 2002, pp. 265–277.

Herzog, Werner, dir. *Encounters at the End of the World*. Discovery Films, 2009.

Howard, John. "The Southern Traverse." *Adventure* (May/June 2001): 31–32.

Howkins, Adrian. "Icy Relations: The Emergence of South American Antarctica during the Second World War." *Polar Record* 42(2) (2006): 153–165.

"How to Retrace the 1912 Race to the South Pole." Text by Joe Robinson; photographs by Geoff Somers: http://www.nationalgeographic.com /adventure/photography/adventure-dreams/south-pole-trek /lessons-learned.html.

Hubbard, Sue (n.d.). "Tacita Dean: At the Tate Britain." http://www .suehubbard.com/sue-hubbard-on-tacita-dean.htm.

Huntford, Roland. *Scott and Amundsen: The Race to the South Pole*. New York: Atheneum, 1984.

Jehlen, Myra. "Archimedes and the Paradox of Feminist Criticism." *Signs* 6(4) (1981): 575–601.

Jorgensen-Dahl, A. and W. Ostreng, eds, *The Antarctic Treaty System in World Politics*. Oslo: Fridtjof Nansen Institute, 1991.

Joyner, C. and E. Theis. *Eagle over the Ice*. Hanover, NH: University Press of New England, 1991.

Johnson, Chalmers. "America's Empire of Bases" (Jan 2004): http://www .globalpolicy.org/empire/intervention/2004/01bases.htm.

Johnson, Nicholas. *Big Dead Place: Inside the Menacing World of Antarctica*. Washington: Feral House, 2006.

Katovich, M. and P. Kincaid. "The Stories in Science Fiction and Social Science: Reading *The Thing* and Other Remakes from Two Eras." *Sociological Quarterly* 36 (1993): 619–639.

Kemp, Norman. *The Conquest of the Antarctic*. New York: Wingate, 1956.

Kern, Stephen. *The Culture of Time and Space, 1880–1918*. Cambridge: Harvard University Press, 1983.

Kirwan, Robert, *The White Road*. London: Hollis & Carter, 1959.

Kitto, Crispin. *Antarctica Cookbook*. New York: St. Martins Press, 1984.

Klein, Naomi. *The Shock Doctrine: The Rise of Disaster Capitalism*. New York: Picador, 2007.

Klotz, Frank G. *America on the Ice: Antarctic Policy Issues*. Washington D.C.: National Defense University Press, 1990.

Kneale, Stephen. " 'You've Got to Be Fucking Kidding!': Knowledge, Belief, and Judgement in Science Fiction." In *Alien Zone: Cultural Theory and Contemporary Science Fiction Cinema*, A. Kuhn, ed. New York and London: Verso, 1990.

Kormo, Fae L. "The Genesis of the International Geophysical Year." *Physics Today* (July 2007): 38–43.

Krapp, Peter. "Between Ectopia and Ecotage: Polar Media." In *Far Fields: Digital Culture, Climate Change, and the Poles*, Jane Marsching and Andrea Polli, eds. Chicago: Intellect Books, 2012, pp. 145–163.

Larson, Edward J. *An Empire of Ice: Scott, Shackleton, and the Heroic Age of Antarctic Science*. New Haven: Yale University Press, 2011.

Leane, Elle. "Locating the Thing: The Antarctic as Alien Space in John W. Campbell's 'Who Goes There?'" *Science Fiction Studies* 32(2) (2005): 225–239.

Le Guin, Ursula. "Heroes." In *Dancing at the Edge of the World*. London: Victor Gollancz, Ltd, 1987, pp. 171–175.

Le Guin, Ursula. "Sur." *The Compass Rose*. New York: St. Martin's Press, 1982, pp. 255–273.

Lenz, William. *The Poetics of the Antarctic: A Study in Nineteenth Century American Cultural Perceptions*. New York and London: Garland, 1995.

Lestringent, Frank. David Faussett, trans. *Mapping the Rennaissance World: The Geographical Imagination in the Age of Discovery*. Berkeley: University of California Press, 1994

Lewis, R. and Karen Wigen. *The Myth of Continents: A Critique of Metageography*. Berkeley: University of California Press, 1997.

Lovecraft, H. P. *At the Mountains of Madness*. 1936. Reprinted. New York: Dell, 1976.

March of the Penguins. Luc Jacquet, dir. National Geographic Films, 2005.

Marsching, Jane and Andrea Polli, eds. *Far Fields: Digital Culture, Climate Change, and the Poles*. Chicago: Intellect Books, 2012.

Marshall, Edison. *Dian of the Lost Land*. New York: Harper Bros., 1934.

Massumi, Brian. *Parables for the Virtual*. Durham: Duke University Press, 2005.

Mason, P. "The Growth of Tourism in Antarctica." *Geography* 85(4) (2000): 358.

Mervis, Jeffery. "A Hot Competition for a Cold Contract." *Science* 325(5936) (July 3, 2009): 20.

Mhyre, Jeffrey D. *The Antarctic Treaty System: Politics, Law, and Diplomacy*. Boulder CO: Westview Press, 1986.

Mickleburgh, Edwin. *Beyond the Frozen Sea*. London: The Bodley Head, 1987.

Miller, Paul. *The Book of Ice*. Mark Batty: New York, 2011.

Mitterling, P. I. *America in the Antarctic to 1840*. Urbana, IL: University of Illinois Press, 1959.

Monty Python. *Scott of the Sahara*, Flying Circus TV Show, episode 23, 1976.

Moore, Jason Kendall. "Bungled Publicity: Little America, Big America, and the Rationale for Non-claimancy, 1946–61." *Polar Record* 40 (2004): 19–30.

Morton, Timothy. *Ecology without Nature: Rethinking Environmental Aesthetics*. Cambridge: Harvard University Press, 2007.

———. *The Ecological Thought*. Cambridge: Harvard University Press, 2010.

Morrell, Margaret and Stephanie Capparell. *Shackleton's Way*. London: Nicholas-Brealey, 2001.

Moss, Sarah. The Frozen Ship: *The Histories and Tales of Polar Exploration*. Katona NY: BlueBridge Books, 2009.

Murray, Carl. "Mapping Terra Antarctica." *Polar Record* 41 (2005): 103–112.

Murray, Carl and J. Jabour. "Independent Expeditions and Antarctic Tourism Policy." *Polar Record* 40 (215): 309–317.

Nasht, Simon. *Last Explorer: Hubert Wilkins, Hero of the Golden Age of Polar Exploration*. NY: Arcade Publishing, 2005.

"Nations Chase Rights to Lucrative Antarctic Resources." *The Epoch Times* http://en.epochtimes.com/tools/printer.asp?id=65009. Accessed 4/16/08.

Naylor, S. et al. "Science, Geopolitics and the Governance of Antarctica." *Nature* (March 2008): 143–145.

Neruda, Pablo. *Las Piedras de Chile*. Buenos Aires: Losada 1961. Translated by Dennis Maloney as *The Stones of Chile*. Buffalo, NY: White Pine, 1987.

Newhouse, Kristina. "Connie Samaras: VALIS: Vast Active Living Intelligence System." *X-tra Contemporary. Arts Quarterly* 10(4) (Summer 2008).

Nielsen, Jerri. *Icebound: A Doctor's Incredible Story of Survival at the South Pole*. New York: Hyperion Books, 2001.

Noble, Anne. (2008). "Artists Statement." http://www.barnard.edu/sfonline /ice/gallery/noble.htm

———. *Ice Blink*. Auckland: Clouds Publishing, 2012.

"NSF Antarctic Logistics." *Science* 325(5936) (July 3, 2009): 20.

Ohio State University Archives. Papers of Admiral Richard E. Byrd, RG 56.1, folder # 2756.

Paglan, Trevor. *Blank Spots on the Map: The Dark Geography of the Pentagon's Secret World*. New York: Penguin, 2009.

Pease, Donald and Amy Kaplan eds. *Cultures of United States Imperialism*. Durham: Duke University Press, 1993.

Pegg, Barry. "Nature and Nation in Popular Scientific Narratives of Polar Exploration." In *The Literature of Science: Perspectives on Popular Scientific Writing*, McRae, Murdo William, ed. Athens and London: The University of Georgia Press, 1993, pp. 213–229.

Peterson, M. J. *Managing the Frozen South: The Creation and Evolution of the Antarctic Treaty System*. Berkeley and LA: The University of California Press, 1988.

Poe, Edgar Allan. *The Narrative of Arthur Gordon Pym*. New York: Harper Bros., 1839.

Pollitt, Katha. *To an Antarctic Traveler*. New York: Knopf, 1982.

Ponting, Herbert. *The Great White South*. London: Gerald Duckworth and Co. [1921], 1999.

Porter, Eliot. *Antarctica*. New York: Dutton, 1978.

Prakash, Gyan. "The Impossibility of Subaltern History." *Nepantla* 1(2) (2000): 287–310.

Pratt, Mary Louise. *Imperial Eyes: Travel Writing and Transculturation*. London and New York: Routledge, 1992.

Pyne, Stephen. *The Ice: A Journey to Antarctica*. Iowa City: University of Iowa Press, 1986.

———. "The Extraterrestrial Earth: Antarctica as Analogue for Space Exploration." *Space Policy* 23 (2007): 147–149.

Ricardou, Jean. "The Peculiar Character of the Water." *Poe Studies* IX(1) (June 1976): 1–9.

Roberts, Susan and Richard Schein. "Earth Shattering: Global Imagery and GIS." In *Ground Truth*, John Pickles, ed. New York: Guilford Press, 2002, pp. 171–192.

Robinson, Kim S. *Antarctica*. New York: Bantam Books, 1997.

Rose, Lisle A. *Explorer: The Life of Richard E. Byrd*. Columbia and London: University of Missouri Press, 2008.

Rosner, Victoria, ed. "Comparative Perspectives Forum: Gender and Polar Studies: Mapping the Terrain." *Signs: A Journal of Women in Culture and Society* 34(3) (2009): 489–494.

Rousseau, G.S. *The Languages of the Psyche: Mind and Body in Enlightenment Thought.* Berkeley and LA: The University of California Press, 1991.

———. *Enlightenment Crossings: Pre- and Post-Modern Discourses, Anthropological.* Manchester UK: Manchester University Press, 1991.

Rubin, Jeff. *The Lonely Planet Travel Survival Kit.* Melbourne: Lonely Planet Publications, 1996.

Sahurie, Emilio. *The International Law of Antarctica.* NY: Springer, 1992.

Salleh, Anna. "Australia's New Antarctic Rights Sparks Exploitation Fears." *Science Online* (April 25, 2008): http://www.abc.net.au/news/2008 -04-25/australias-new-antarctic-rights-sparks/2415836

Samaras, Connie. "American Dreams." *Scholar and Feminist Gender On Ice Special Issue* 7(1) (November 2008): http://www.barnard.edu/sfonline /ice/samaras_01.htm.

Scott, Joan. *Feminism and History.* Oxford and New York: Oxford University Press, 1996.

Scott, Joan. "Gender: A Useful Category of Historical Analysis." *American Historical Review* 91(5) (1986): 1053–1075.

Scott, K. "Institutional Developments within the Antarctic Treaty System." *The International and Comparative Law Quarterly* 52 (2003): 473.

Scott, Robert Falcon. *Scott's Last Expedition: The Personal Journals of Captain R. F. Scott, C.V.O., R.N.* London: Smith, Elder, and Co., 1913.

———. *The Voyage of the Discovery.* London: Smith, Elder, and Co., 1905.

Shackleton, Ernest. *South.* London: William Heinemann, 1919.

Shapiro, Deborah and Rolf Bjelke. *Time on Ice: A Winter Voyage to Antarctica.* Camden ME: International Marine, 1998.

Shepheard, Paul. *The Cultivated Wilderness, or What is Landscape?* Cambridge and London: The MIT Press, 1997.

Simpson-Housley, Paul. *Antarctica: Exploration, Perception, Metaphor.* London and New York: Routledge, 1992.

Smithson, Robert. *Collected Writings.* Jack Flam, ed. Berkeley: University of California Press, 1996.

Sobieszek, Robert. "Robert Smithson's Proposal for a Monument in Antarctica." In *Robert Smithson*, Eugenie Tsai et al., eds. Los Angeles: The Museum of Contemporary Art, 2004, pp. 142–147.

Solomon, Susan. *The Coldest March: Scott's Fatal Antarctic Expedition.* New Haven and London: Yale University Press, 2001.

Spillers, J. "Re-imagining United States Antarctic Research as a Defining Endeavor of a Deserving World Leader: 1957–199," *Public Understanding of Science* 13 (2004): 31–53.

Springer, Claudia. *Electronic Eros: Bodies and Desire in the Postindustrial Age.* Austin: The University of Texas Press, 1996.

Spufford, Francis. *I May be Some Time: Ice and the English Imagination.* London and Boston: Faber and Faber, 1996.

Spufford, Francis. "On Observation Hill." *Granta* 67 (2005): 112–119.

Standish, David. *Hollow Earth: The Long and Curious History of Imagining Strange Lands, Fantastical Creatures, Advanced Civilizations, and Marvelous Machines Below the Earth's Surface.* Cambridge, Mass: Da Capo Press, 2006.

Stokke, O. and D. Vidas, eds. *Governing the Antarctic: The Effectiveness and Legitimacy of the Antarctic Treaty System.* Oxford: Cambridge University Press, 1996.

Stone, Allucquère Rosanne. *The War of Desire and Technology at the Close of the Mechanical Age.* Cambridge MA: The MIT Press, 1996.

Suter, Keith. *Antarctica: Private Property or Public Heritage?* London: Pluto Press, 1991.

Szeman, Imre. "The Rhetoric of Culture: Some Notes on Magazines, Canadian Culture, and Globalization." In *Cultural Studies: An Anthology,* ed. Michael Ryan, 82–98. Oxford: Blackwell, 2008.

———. "System Failure: Oil, Futurity, and the Anticipation of Disaster." *South Atlantic Quarterly* (Fall 2007): 805–823.

Templeton, M. *A Wise Adventure: New Zealand and Antarctica 1920–1960,* Wellington: Victoria University Press, 2000.

Thompson, Natto. *Experimental Geography: Radical Approaches to Landscape, Cartography and Urbanism.* New York: Melville House, 2009.

The Truman Show. Peter Weir, dir. Paramount, 1998.

Tsai, Eugenie et al. *Robert Smithson.* Los Angeles: The Museum of Contemporary Art, 2004.

Turchetti, Simone, et al. "On Thick Ice: Scientific Internationalism and Antarctic Affairs." *History and Technology* 24(4) (Dec 2008): 35–376.

Verne, Jules. *Le Sphinx des Glaces.* Paris: Pierre-Jules Hetzel, 1897.

von Uexkull, Jakob. "A Stroll Through the Worlds of Animals and Men: A Picture Book of Invisible Worlds." *Instinctive Behavior: The Development of a Modern Concept,* ed. and trans. Claire H. Schiller. New York: International Universities Press, Inc., 1957, pp. 5–80

Wiley, John. "Becoming Icy: Scott and Amundsen's South Polar Voyages 1910–1913,' *Cultural Geographies* 9 (2004): 249–265.

Wilkes, Owen and Robert Mann. "The Story of Nukey Poo." *Bulletin of the Atomic Scientists* 34 (October 1978): 32–36.

Wolfe, Tom. "The Doctrine That Never Died." Op-ed. *The New York Times.* January 30, 2005.

Yusoff, Kathryn. "Visualizing Antarctica as a Place in Time." *Space and Culture* 8 (2005): 381–398.

———. "Antarctic Exposures: Archives of the Feeling Body." *Cultural Geographies* 14 (2007): 211–233.

Zengerie, Jason. "More Heat, Less Light." *New York Magazine* (June 2010): http://nymag.com/guides/summer/2010/66795/.

Ziskind, Mariano. "Captain Cook and the Discovery of Antartica's Modern Specificity: Towards a Critique of Globalism." *Comparative Literature Studies* 42(1) (2005): 1–23.

Index